The Forgotten Man of

CW00449347

THE FORGOTTI
LAKELAɴᴅ

the story of

WILLIAM T PALMER FRGS MBOU FSA[Scot]

by

SHEILA RICHARDSON

Mill Field Publications
1997

ISBN 0 9526665 4 5

Published by
MILL FIELD PUBLICATIONS
10 Seadown Drive, High Harrington,Workington, Cumbria
CA14 4NE

Front cover photograph
S.J.Studios, Workington

Printed by
Derwent Press, Workington

Contents

Acknowledgements

It would have been impossible to tell the story of William T Palmer without the help of many individuals and organisations. In this respect, I am deeply indebted to the following;

Ruth Buchanan, John Chesterton, Brian Palmer, Jack Palmer, Jack Walker, the Governors and Headmaster of St Oswald's C of E School, Burneside, the staff of the County Records Office Kendal, The Fell and Rock Climbing Club, the Huntington's Disease Association, Lancashire Evening Post, City of Liverpool Record Office, News International plc, Royal Geographic Society, Helena Thompson Museum, Workington, Cyclist Touring Club, Liverpool Daily Post and Echo, and most of all to Christine Buchanan, the grand daughter of William T Palmer who gave me free access to family photographs and documentation.

Introduction

I first came across the name of William T Palmer when researching for an anthology, "A Lakes Christmas." Part of the brief from the publisher was to search out any previously published interesting and unusual pieces that referred to the Lake District between the 1st of December and Twelfth Night.

Many a library shelf was scoured, and depths of archive material investigated in the search. One writer's name that kept emerging was that of William T Palmer. In the local reference section of most of the Cumbrian libraries, there were whole shelves of his books about the Lake District.

Who was he? What was he? It was a name with which I was totally unfamiliar.

Librarians, and others with a wide knowledge of Lake District writers were asked for any information about the man. In each case blank looks greeted the queries.

"W T Palmer?"

"Never heard of him" was the usual response.

After a brief scanning through many of his 40 books, I was little the wiser as to what kind of man he was. The publication dates of the books ranged from 1903 until 1952, and they gave a nostalgic insight into what the Lake District was like in less frantic times than those in which we live in today. They were rich in folk lore and custom; they contained detailed accounts of foxhunts in wild valleys; he took his reader to the high fell tops, and shared with them the secrets of the valleys.

Some of the subject matter in Palmer's books indicated that he was a competent mountaineer; an original member of the Fell and Rock Club, and a former editor of its Journal.

The title pages of his books increasingly showed letters appended to his name; first of all FRGS; a few years later MBOU; and finally FSA[Scot].

How could a man who seemed to have made such a contribution to his native county in the fields of literature, natural history, and mountaineering remain virtually unknown? Unknown not only to my generation who explored the Lake District from the 1950's onward, but to those of an older generation, whose age and interests, overlapped the Palmer era.

Although his books were on the library shelves in plenty, there was nothing entered under the name of W T Palmer in authors' directories or publishing year books for the period of his writing career. Apart from a number of dusty volumes on many a library shelf, there was nothing to mark the 77 years of his existence.

During his lifetime, which occupied the last twenty years of the nineteenth century, and the first half of the twentieth, there was no mass means of communication to promote his work. There was no radio or television space to give interviews that advertised his latest book. Yet in spite of this a number of his books went into many reprints and updated editions. In those years, relatively few visitors, compared to today's millions, came to the Lake District, so there were not the numbers of interested readers to enjoy his books. They went out of print and out of fashion before they could be appreciated by a wider national and international body of readers. A family friend of William Palmer claimed that "he was a man ahead of his time."

Yet his books probably document the social history, the way of life of rural communities, and express an appreciation of the Lake District landscape better than any other writer. The style of his writing may not have the purity and romanticism of the Wordsworths; it may not have the scholarship of Ruskin or Rawnsley, but the range and variety of his subject material can hardly be surpassed by any other writer.

The subject material of his books was not only restricted to his native county of Westmorland, and its Cumbrian neighbour. He wrote, with authority, about the history, people, and landscape of other parts of England, as well as of Scotland and Wales. His books are readable and informative, and although it is over ninety years since he put pen to paper to produce his first book, much of the content is still of interest today. While the march of technological progress has undoubtedly obliterated much of Palmer's world; his words take us back to a less frantic age of solid values and ideals. They, above all else, provide a link with our heritage, be we English, Welsh or Scots.

Because there seemed to be hardly any documentation about the man, I felt that his literary contribution should be publicly acknowledged, and the name of William T Palmer receive some of the credit that it justly deserves. But where to begin?

Many influences come together, in various degrees of importance, to combine to shape a life; and to change a person into a personality. In attitudes and emotions we reflect our association and interaction with others. Enthusiasms develop through involvement with interests and activities. Combined together, they weave a web of physical and mental growth that is, inevitably, individualistic.

The ingredients for life may be the same in many cases, but the quantities in which they are weighed, differ with every one. The pattern of a person's life is a kaleidoscope of genetic inheritance, which although initially laid down in a basic array of colours, is constantly reshaped as experiences, circumstances and reactions change their design. A brash youth of 17, while carrying the same identity throughout his lifetime, is in many ways a different man to the sage of 71.

How then to discover William Thomas Palmer?

There are few people alive today who have a personal recollection of the man, but those who do, all feel that his story has laid hidden for too long. "He never had the recognition he deserved" is the comment of them all.

Thus began a personal quest for information about the man. A chance inquiry of some ladies leaving the church in his home village of Burneside, gave the information that there were still some relatives of William Palmer living there. They were contacted and provided further leads to other family members who were able to supply more information.

My first intention was purely to satisfy an initial, and personal curiosity about the man. But the information kept accumulating, and with the acquisition of each new piece, there was the feeling of completing a huge jigsaw of a person's life and personality. Insignificant sounding fragments, often of little informative value in their own right, slotted into place when set against the context of other pieces of the puzzle.

William Thomas Palmer

The framework of the jigsaw, some might call the straight edged pieces, were William Palmer's own books. They had to be sifted through for gleanings of information about his personal life, his family, his philosophy and some of his many accomplishments. Blocks of the pattern were filled in by researching into archive material, newspapers, journals of the Fell and Rock Club, and letter writing to any number of individuals or organisations that had connections with William Palmer at some time during the course of his life. Meetings and interviews with direct family members and former friends added to the information that began to trickle in. The shadowy outline of the Lake District's forgotten man changed, as he began to take on a substance and personality. Family photographs were made available, that actually showed what he looked like.

He appeared to be a sturdy, resolute individual. His body posture denoted self confidence and strength of character. He looked at ease in front of the camera, confident and self assured, yet without any arrogance. His straightforward gaze into the camera lens was of honesty and dependability. I felt that here was a man I should like to have known. Instead, I met with his only surviving direct relative, a granddaughter slightly older than myself. Her facial resemblance was uncannily like that depicted in the photographs of William Palmer.

It is almost forty years since her grandfather died, a length of time during which memories of loved ones, experiences and events can fade into shadow, especially if they took place in one's early life. Over that period, the memory of her grandfather had paled. His achievements had dimmed, through lack of recognition. She had parted with all his books save one; a smoke damaged copy of The English Lakes. On the fly leaf was written an inscription. "To my wife, from the Author." On the same page was an impressed and dried four leafed clover .

Two days after my initial visit to Christine Buchanan, I received a letter of thanks, expressing her gratitude at re-introducing her to her grandfather.

I hope that this biography will introduce many more to the life, writings and achievements of William Thomas Palmer, who in spite of all he did has remained, for many years, THE FORGOTTEN MAN OF LAKELAND.

Chapter 1 - CHILDHOOD

William Thomas Palmer was born on the 8th July 1877 in his parents' home, at 4 Bowston Terrace, Bowston, which is almost an extension of the larger, and neighbouring village of Burneside, near to Kendal. He had two older brothers and two sisters, one of whom was younger by two years, than himself.

Bowstone is a hamlet rather than a village, and in later years when William Palmer was writing about his childhood days, he used the spelling " Boustone", which was the local pronunciation. Modern day maps however, use "Bowston", for the hamlet that is situated on a quiet back road just a few miles north of Kendal.

The house in which William was born was built of stone, a grey cottage that stands alongside the road that once carried the horse drawn coach traffic from Kendal to Staveley. Some of William's earliest memories were of awakening in the morning to the approaching sounds of pounding hooves, or the jingle and creak of harness. Outside, the swirling dust and the steamy smell of sweating horses were accompaniments to the quavering high pitched resonance from the coachman's "yard of tin", as it blared a warning of the galloping team on the narrow roads.

The Palmer's cottage was small, yet it was sturdily, if simply built with two rooms upstairs, two rooms downstairs and no other refinements that are take for granted today. In common with most other rural cottages of that time, an earth closet at the bottom of the garden was all that was available in the way of sanitary facilities. The inhabitants of these cottages accepted as normal, what would now be regarded as primitive conditions; for a shovelling of cinders or soil was a forerunner to the flushing of a WC.

William's mother, Jane, was two years older than her husband James. They both were of country stock and each came from a family background that had had close association with farming and country life through many generations. William Palmer's paternal great grandfather was a drover, who with others, used to drive large numbers of sheep and cattle on the long slow journey between the Shap area of Westmorland, and over the Border into Scotland. Some of his journeys took him far to the north, into the sparsely populated areas of the highland regions. He travelled along many of the green or drove roads, and a number of the routes that they used, can still be followed today.

Much of the countryside through which the drovers journeyed was wild and lonely. The duration of their journey depended on the speed of movement of their animals, for they had to be fed, watered and rested on the way. If they were driven too hard, the beasts lost weight and their value was subsequently reduced. The drovers then had the reckoning to face.

Frequently, the weather conditions that they encountered were as harsh as the countryside through which they travelled. It was on one such journey that William Palmer's great grandfather lost his life, while trying to rescue sheep that were buried in the snow during a severe blizzard in the Shap area. "At least that's the tale that has been

No.4 Old Bowstone -
the house in which William Palmer was born

The former Burneside Board School,
now part of James Cropper plc.

handed down through our family for a number of generations," explained a nephew of William Palmer who lives in the village of Burneside.

During the whole of William Palmer's lifetime, Bowston was in the county of Westmorland, and remained so until the national reorganisation of the counties in 1974, when Westmorland was joined with Cumberland to become the larger county of Cumbria. William Palmer was a Westmorland man at heart and proud of his county's heritage. Although in later life he was to travel extensively throughout many different parts of the country, he never forgot his Westmerian roots.

Some of the earliest documented information about William Palmer can be gained from the 1881 Kendal Census records. It gives an insight into some of the Palmer family details. His father was described as a labourer, aged 29 of Bowston Bridge. At that time, the term "labourer" carried more prestige than it does today, for it indicated that such a man was capable of turning his hands to all manner of tasks. These included dry stone wall building, wood and coppice management, stock rearing, shepherding, to name but a few. Much of James Palmer's early working life was spent on farms in the surrounding district, where he worked mainly with sheep. It was within this background that the young William grew up. It gave him a lifelong interest in sheep and an affinity with shepherds that made him completely at ease in their company.

William's mother was Jane, aged 31, at the time of the Census. There were five children in the family; John, 9; Jane Elizabeth [usually known as Bessie] 8; James Edward 5; William Thomas 3; and Mary [May] 3 months. In the census returns, John, Bessie and James were described as scholars, a term that was not applicable to William until two years later.

Family memories indicate that the years of William's childhood were happily spent, even though there was little money available. The cramped and limited cottage accommodation was shared, for most of the time by the seven members of the family, together with a collie dog. "They were never without a dog, but it was always his mother's dog," a member of the family recalled.

The house, that still stands today, has a high gabled roof and forms part of a terrace of four trim cottages of grey stone. They now look more suited to holiday accommodation, than the homes of growing families. Only a tiny front garden separates the house from the quiet road that leads to Kendal. Parallel to this road, but far removed by the intervention of green fields, is the modern connecting road from the M6 motorway, where the rush of tourist traffic to the Lake District hurtles by.

It was at Number 4 that William Palmer spent his childhood years, and from there, grew into early manhood. The fields and river behind the house formed his playground, the woods were places of adventure, and the ivy festooned Victorian post-box at the end of the row that is still in use, was the recipient of his early letters to boyhood heroes.

The Palmer home was about a mile from the nearest school in the neighbouring and larger village of Burneside. William attended as a full time pupil, from the 29/8/1882, until he left formal education on 30/9/1892, at the age of 15. The school was typical of many of the Victorian structures that spread throughout the country in the late 19th century. It was solidly built of stone, and its character forming curriculum and the strict disciplinary regime to which the scholars were subjected, were as substantial as the school's foundations. Although the old school building still exists today, it is not as an educational establishment, but as part of the administrative block of a paper mill. A spacious modern building has now replaced the old school.

When William first started to attend school, his parents were aware that "those who attend the school must pay their weekly fee and also obey the rules." For those that didn't, punishment was swift and harsh; it was not unusual for children to receive, "3 strokes across the shoulders for turning sulky."

A number of parents also contributed to a Clothing Fund which was organised by the school's managers, from which regular pay outs were made during the school year. Once a parent had contracted to pay, they were expected to send in regular contributions. This was not always the case and one mother frequently complained about her children "being sent home for the school pence and clothing money."

One of the managers of the school was local business man and Kendal 's Member of Parliament , James Cropper, the owner of the paper mill in the village. It was at his suggestion, which is recorded in the School Log Book on the 11/8/1884 that, "He desired the attention of the parents to be called to a scheme for supplying a dinner during the cold months to each child attending school. The cost 1d per day. Cooking, room, attendance, free of cost." This scheme was implemented on 3/11/1884 when the first penny dinner was served in the school. William was then aged seven and was still in the infants department. It is recorded that the first penny dinner was not only attended by the children, but also by James Cropper MP, Mrs Cropper, Miss Cropper, Mrs Jones the Master's wife, and Mrs Bryce, the wife of a benefactor of the school. There is no record of what the meal consisted, but on a later occasion, on the 18/1/1887 when there was thick snow on the ground, the meal was recorded as rice milk.

The school was normally closed to the pupils at lunch time, but on one recorded occasion, 14/9/1883, "a village boy entered school during dinner hour, and with chalk wrote on a slate some filthy words." The headmaster contacted the boy's parents, "without any hope of the boy being taught the sinfulness of his ways."

Attitudes to children starting school in the latter years of the 19th century, were different than those of present times. The fact that parents had to pay the school pence when their children attended ensured that they wanted to see some results for their money. The Master expected that the children would "know their letters" before being enrolled in the infants department of the school. There are a number of entries in the Burneside school log book, deploring the fact that infants were starting school WITHOUT knowing their letters.

William Palmer's achievements in later life are all the more remarkable when set against the background of his elementary village school education. His ten educational years of Board School experience were at the hands of only two certificated teachers, who were assisted by a sprinkling of pupil teachers. Many of these were just older pupils of the school, who hoped to gain qualified status after a number of years of training, that incorporated the practical, "hands on", experience.

The headmaster of William's school was Thomas Jones. He was the certificated teacher in charge of the First Class, which was the senior class in the all age school. During William's schooldays, the number on roll varied between 100 and 120 in the boys and girls department, and between 40 and 50 in the Infants department. The teaching of about 150 children, the majority of whom attended on a full time basis, was shared by a Headmaster, two assistants and two pupil teachers. The certificated teachers, of which there were three at times, had the dual responsibility of not only teaching the children that were currently on roll, but of supervising and training the pupil teachers. "A pupil teacher was reported as unfit for work by playing cricket until the last minute." 28/4/1883. Some four months later, "the pupil teacher was late for morning lessons, none given. Employed during first hour as a punishment."

These then were the people in whose hands William spent his formative years in the Mixed Infant department of the school, which was overseen by only one qualified teacher.

Mixed Infants, was a common term of reference for those children below the age of eight. There was little difference made between the needs of boys and girls, until they had left the infant class behind, to move into the "big school." Although the complete age range of children in the school was from five to fifteen, most children left school when aged between thirteen and fourteen.

The scholars were divided into two classes. The first class of older children was taught by the headmaster who sometimes had the help of a qualified assistant, while the second class was taught by an assistant certificated mistress. The number of children in each class was inevitably high, and it was not uncommon for a teacher to have a class of over fifty children. The first class was further sub-divided into seven standards, some of which were under the supervision of pupil teachers.

There was a constant demand from the village paper mill to employ children as "half timers." These "half timers" were drawn from the senior end of the school, and caused the headmaster considerable problems in trying to complete the requirements of the curriculum. 23/11/1883 "The Mill wants so many half timers that the master finds great difficulty in keeping a creditable standard VI and VII." This half time attendance by a large number of children caused considerable disruption not only to their own learning opportunities, but to the general run of school life, for one of the major difficulties that the Master had to contend with was time tabling lessons for large numbers of a wide age range with limited staffing resources.

Over the ten years in which William attended Burneside School, there were also many family upheavals in the village as the fortunes of local folk fluctuated. Many of the families were very poor and they left the village to seek work and greater prosperity in nearby larger towns. They invariably took their children with them which caused the school population to vary enormously as most of the families had four or five children in attendance. This created all sorts of problems in the organisation of the "standards", and affected the number of teachers that were appointed to the school. The teachers task in ensuring continuity of progress in their large classes of mixed age and ability, was made the more difficult as the attendance of many of the children was very irregular. There were a number of absences caused through the "severity of the weather", ill health, and prolonged colds, or truancy as well as half time in the Mill or the harvest fields. These interruptions to the weekly round were borne with great stoicism by the head teacher with frequent entries in the log book of, "work going on as usual." Such uncertainty had a disrupting effect on the school's programme, in spite of the sterling efforts of the Master to maintain a degree of stability in curriculum and organisation.

Throughout his schooldays, William must have been a middle of the road sort of character, for if there are no records of outstanding achievement, neither are there any entries recording frequent absences or punishments that relate to him. Some scholars however, were brought to book; "[one boy] was punished for gross disobedience", and "two boys accused of stealing duck eggs were referred to their parents." As well as administering a few strokes of the cane to any miscreant, the school master had his own way of enforcing discipline. This was by gripping and pulling a lad's hair, which William Palmer later recalled "made the stoutest lad shrink." In spite of the rough justice that was meted out, William bore no hard feelings towards the schoolmaster, even though he had on occasion received a swish or two from the cane. In later years he reflected, "He made a fair job out of some rather wild youngsters."

The headmaster was well aware of the problems he had to deal with, for the school entry in the log for 29/7/1889 recorded that with the new intake of children, "thirty were very rough in manners. Put each class through a course of drill." He put it down to the fact that "discipline suffered through the thoughtlessness of the pupil teachers, and the worrying propensities of the assistant."

The location of the school, which was built beside the river Kent, which makes its journey down from the high grounds of Kentmere, was disadvantageous to the teaching staff. The main school buildings were connected to the playing fields on the other side of the river by a bridge. In the right conditions of weather and season, especially when the fish were running, the river provided a more attractive and alternative distraction to many of the lads, who preferred fishing to lessons.

One lad "got belted in school when he whispered too audibly that he wasn't going to stop in after school." He knew that there were fish running up the river. He also knew where the postman had hidden his rod while on his rounds earlier that morning.

The river provided other distractions for the Burneside scholars, one of which was heralded by the yelping pack of Kendal otter hounds. William was a keen follower of these dogs who used to hunt their prey all the way up the beck, as far as the valley of Kentmere. The splashing of the hunt followers and the baying of the hounds was a greater attraction to young lads, than the schoolmaster's bell that summoned them to learn English grammar or complicated mathematical processes.

The frustration of an overworked and underpaid head of a small village school creeps into his log book entry of 13/11/1885; "Standard 1 are still very bad in dictation - they don't seem to improve at all"; this was William's class .

Each year, the school was subjected to an inspection to ascertain that it re - qualified for the amount of grant money to which it was entitled. In those days, payment for the running of the school was based on the results achieved by the scholars. The school year was divided up into quarters and each child was expected to attain the necessary standard of work that was pertinent to each quarter. If that standard was not achieved, then the child was not allowed to progress to the next stage of his education. However, even in those far off days, the fact that some children had learning difficulties was recognised, for some of the pupil teachers spent their time "bringing on the dull children."

Many of the summaries of reports submitted by Her Majesty's Inspector, [Victoria was Queen when William attended school], and recorded in the school log books, indicate some of the weaknesses and successes of the old Board school system. The reports also gave the opportunity to mark the progress of the pupil teachers. Some had particular difficulty in satisfying the Inspector and school managers of their ability to teach and keep control of their charges. In some cases, their aim of attaining the recognition of qualified status, was only achieved after many years of effort. The annual school inspection reports indicated that the infants were "kindly managed", although it was felt that there should be more variety in "their occupations." Arithmetic was good, but reading and spelling were far from satisfactory in William's standard.

Singing and Drill were recorded as being excellent. Drill involved the children performing a series of exercises to the teacher's command where William took his place in a row, among the regimented lines of children. Discipline was strict, for each movement had to be performed in unison and in exactly the same manner as the rest of the class. There was trouble for anyone who was out of synchronisation with the rest of the group. In some ways it was like the performance of a military unit on the parade ground; only the size of the participants, and the manner of dress was different.

No distinguishing uniform was worn by the children that attended Burneside School during William's schooldays. Money for most families was in short supply, though the regular contributions to the Clothing Fund ensured that money was available through repayments at Christmas and Easter. In some respects, the Clothing Fund served the purpose now adopted by mail order catalogues , where payment is made by instalments. The difference in William Palmer's day was that the money was saved up, then clothing bought, which is the reverse of modern day procedures. The general practice was that families were very careful in looking after their clothes. They were darned and patched, and as one child grew out of a size, it was handed down to other members of the family or passed on to relatives, or any one else that it would fit. Often, the gender of the wearer was immaterial to the garment that was available; fashion counted for nothing; anything that fitted was worn, although sometimes the "fit" left something to be desired. School clothes were functional and serviceable; they were often home made and sometimes even home spun. In the late nineteenth century, Kendal was noted as being a woollen town, where the manufacture of woollen goods went alongside the many hat makers, shoemakers, cloggers, tailors and dressmakers, glovers and even stay makers.

There was a tradition among the village folk for clothes making, repairing, patching; darning , mending, and it was clad in practical, tough, everyday clothes that William went to school. The same clothes were also worn at home and for any farm work that he did after school or at weekends and holidays. He did have an outfit that was kept "for best", but the ownership of even this changed as he grew in size. In many rural areas, some families followed the tradition of children being stitched into their underclothes for the winter. Although this practice survived in parts of Cumbria well into the 20th century, there is no evidence that William was ever subjected to that indignity.

In the first two years that he attended school, his clothes were still those of a very young child. This was common practice, for along with other boys of his age, he wore a pinafore "dress." This was normal attire for both boys and girls in those Victorian days, for boys did not wear trousers until they reached the age of seven. When that momentous occasion came about, they were deemed to be "breeched."

The school which he attended every day apart from the six weeks of holidays, could hardly be described as a comfortable place of learning. High ceilings sucked away the little warmth that was spread from the open fires that heated the classrooms during the winter months. The children took any, and every opportunity to cluster round the flames; toasting alternate body sides to warm themselves on the cold, dark days of winter. The opportunity to stand at the teacher's desk was thankfully taken, whatever the reason. Those not so fortunate, shivered.

The classrooms were filled with rows of close packed, iron framed desks; their dark oak a distant memory of early growth in a sunlit forest. The cramped conditions the desks afforded, ensured there was little room for individual movement.

Hanging from the ceiling were spluttering yellow gas lamps. Their feeble light, controlled by dangling chains, caused heads to hover inches above a book, as eyes followed in unison along a close typed page.

Chalks squeaked on slates; steel nibs scraped and sputtered ink across a page, where careless blots invoked a teacher's wrath. Life in the crowded classroom was far from comfortable, and this fact was noted at one of the annual inspections, when recommendations for improvement of conditions were made. The following year, an inspector's report indicated that "the schoolroom had been considerably improved by the removal of a row of desks," although the "floor should be boarded." There was a plea, in one Inspector's report, to make the Gallery a "less uncomfortable place for little children." This indicated that the Inspectors had the interests of the children at heart, if not, apparently, those of the long suffering teachers who repeatedly came in for criticism.

The HMI report of 1886, when William was aged 9 and in Standard 2, showed that while the teaching of history and geography was extremely well done, "the more important elementary work is less than satisfactory." William's later interest in geography and geology, that is evident in his published books and articles, could well stem from the fine teaching he received in these subjects, at an early age.

When William Palmer was well on to middle age, he was engaged in collecting material for his book the "Verge of the Scottish Highlands." The sound and enthusiastic teaching that he had received in history, when at school was acknowledged in the following way.

When William Palmer was well on to middle age, he was engaged in collecting material for his book the "Verge of the Scottish Highlands." The sound and enthusiastic teaching that he had received in history, when at school was acknowledged in the following way.

"At Stirling Bridge I had a vision.Why did my mind leap back a few years to school histories, and fix the place as the scene of the battle in which William Wallace so thoroughly smashed an Anglo Norman force."

William's teacher must have made the Scottish hero come alive to his young pupil, in a way that ensured the dormant seeds of memory would ripen into full bloom when the time arose.

In 1887 when he was aged 10, and in Standard 3, William's class escaped both praise or retribution in the annual inspector's report. This was in spite of the fact that the corner-stones of the future writer's craft, namely, reading, spelling and grammar all appear to have been poorly taught throughout the rest of the school. Although William's class escaped criticism for that year, he was still being taught by the same teachers who had so displeased the inspectorate.

In 1888, the inspector appeared to be losing patience that standards were still not improving, and indicated, "Better results in written arithmetic and spelling will be looked for if the present rate of merit grant is to be again recommended." William was now in Standard 4. where part of his curriculum included the recitation of poetry, "Casabianca", and chunks of Shakespeare; this came in for great praise from the inspector. The training of his memory in this way, later proved to stand William Palmer in good stead, when recalling tales passed on to him by local folk.

Schooling encouraged William's enthusiasm for reading, although it was narrowly confined to the books of the curriculum that included such classics as Tom Brown's Schooldays, or Ivanhoe. The reading of other material was positively discouraged. In his Log Book, the Master made the following entry; "A boy found reading a worthless and immoral book. Punished the boy and admonished the whole school."

In spite of this admonishment, William Palmer was an avid reader of any book that he could get hold of, although he admitted that he did not fully understand Wordsworth until he was turned 18. This keenness set him apart from the village lads, for there was a certain apprehension among the country folk of his village that the ownership of a few books led to an unhealthy reputation. They were somewhat in awe of people with "a lot of learning" and it did not seem right that one of their own should join those ranks. Anyone who did, was regarded as definitely odd.

There was one such character in the village who was noted for having a large collection of books. He was referred to as a "wise man", the title was conveyed in jest, rather than the respect that wisdom often brings. He earned the title only because he possessed the books, not through his ability to read them. William Palmer longed to get his hands on those books , much to the dismay of his "good grandmother" who was afraid that through reading them, he would be "turned into a pillar of salt or be consumed by fire." The same "wise man," from whom William did eventually borrow the books, also had a horror of "half eddicated lads as set theirsels up to Kna' summat." Nevertheless, he allowed the young Palmer access to his books which were borrowed whenever he wanted.

Many years later,William shared more memories of his school days with some other children he encountered while walking between Dulsie Bridge and Cawdor during one of his many visits to Scotland. He asked the elder of the two girls which class she was in.

"The fourth"

William told her, " in my village school in England, we began to take real grammar lessons in Standard 4, and practice [multiplying and dividing £sd] in Standard 5."

When the girl told him that she would be attending secondary school at Nairn, his response had overtones of regret, "I said that at my school we had no chance of secondary or university education, but the present children had."

Over fifty years earlier, in 1889, when William was aged 12 and in standard 5, the Inspector's report began, "Inaccuracy is the weak point and intelligence the strong point." In an arithmetic test, William's standard achieved the highest percentage of correct answers with 49%; while the oldest children in the school who were in standard seven only achieved 9%. This brought the directive that "Arithmetic must improve in the upper standards if a merit grant corresponding to the present, is to be maintained."

These were severe words indeed. The teachers' rates of pay were placed under threat, to say nothing of the funding of the school if the standard of performance among the children did not improve. At that same inspection, a pupil teacher who had tried for a number of years to achieve a pass level to qualified status, finally succeeded, but was told to "attend to method."

Part of the Burneside school curriculum, included some preparation for the practical arithmetical calculations involved in farming and land management, that boys would encounter when they left school. Their headmaster, Mr Jones ensured specific and practical tasks were incorporated into the arithmetic curriculum.

"Every juvenile farmer was expected to figure a bit" in order to handle accounts, order supplies, and estimate yardages for fencing, walling and coppicing. The ability to handle mechanical arithmetic processes was vital to their everyday livelihood. The lads were set problems about the village, for which they were equipped with lengths of knotted string for measuring, and slates and pencils for recording. Even such difficult tasks as working out how much straw was contained in a stack with a conical roof was not expected to be beyond their capabilities. Their results proved to be remarkably accurate in spite of some good humoured, but derogatory banter from the village men. Calculations of complex measurements of woodland, or wall length involved working by the rood, or fall [about 22 square yards]; these were done without any of the mechanical or electronic aids that are available today, and the task of measuring a field involved weary hours of calculation. "The job had to be finished ere we were allowed to go home."

Yet in spite of the rigorous routine of work and discipline, there were plenty of half day holidays given to break up the humdrum pattern of school life; but some were only achieved with noisy persuasion. It was a tradition to enliven the day before Good Friday with the clattering of broken buckets, pans and kettles. These were collected from around the village, tied to a rope and dragged along the road to school, accompanied by the chorus of "Trot her'in; trot horn; Good Friday tomorn." This was an indication to all and sundry, but especially the Master, that after a rowdy morning spent in school, the afternoon would be the start of the Easter holiday.

It is still traditional in the Lakes counties to have hard boiled eggs at Easter. These can either be dyed with the natural colours that come from onion skins, or gorse; or can be hand painted or decorated. One tradition that has almost died out is that of rolling the eggs and this often took place on Easter Monday, when families would gather at a local hilly spot, to roll their hard boiled eggs down a slope. The winner was the one whose egg did not crack on the way down.

Another Easter tradition in which the young William Palmer joined, was to tour the neighbouring farms and hamlets in the company of the Pace Eggers. This usually consisted of a group of young adults, or lads and lasses who dressed themselves up as traditional characters to perform mumming plays. The theme of these productions was always the triumph of Good over Evil, although some of the Good characters had narrow escapes before eventually winning through.

One of William's earliest childhood memories of the Pace Eggers was of being allowed to stay up late to watch the play, but in later years, when he was old enough to join in, he found that taking part could be a hazardous affair. Farm dogs were not able to distinguish between Good and Evil, and did not always take kindly to the sooty, sack clad characters that invaded their territory. "The sheep dogs wherever we went used to harry us - figures with black faces and streamers of coloured paper came not within their philosophy."

But with the reward of fresh eggs to take home, the discomfort of few bruises from the mock fights and frantic chases by barking dogs, could be tolerated. William recalled these experiences in his first book, "Lake Country Rambles" where he also recorded the dialogue and the characters that took part in the traditional performance. It is a valuable record of part of the once common Easter festivities in the Lake District, for there is little documentation elsewhere of this annual, but now almost forgotten event.

Easter Sunday was the occasion when it was customary for all the children to have some new item of clothing to wear as they paraded through the village on their way to church. In William's case, sometimes all that could be mustered was a new tie that he shared with one of his brothers.

The Burneside village church of St. Oswald' s was closely involved with the school activities, as it continues to do so today. During William Palmer's schooldays, the vicar of the church also had the same close liason with the school in his role as one of the managers. He made regular visits to ensure that the children were being correctly taught, especially in religion and moral values.

A popular event in the village was the annual band festival, that was held to celebrate St. Oswald's Day. This was always well received by the scholars, for it gave them the opportunity of a half day holiday. Another annual event of significance to village and school life, was "Confirmation Day", for this had the tradition of awarding of a holiday for that occasion, which was much appreciated by the children.

Less unlikely opportunities for a holiday, were the very rare times that the headmaster was ill. With limited staffing resources, large classes of mixed ability and age, a demanding inspectorate and managers, one would have sympathised with a master who sought the relief of a half day break. But Mr Jones was a man of character and obvious good health, for few occasions were recorded during William's ten years at school, when the Master was absent.

To supplement the legitimate holidays that were granted, many of the scholars were not above taking their own time off when they felt the need arose, for "A circus in Kendal was the cause of many absences" was recorded in the school log book.

When the four weeks of the summer holiday arrived, William was often sent to spend part of the time in the Levens area of Lancashire, where his mother had been born, "on the edge of the mosses next to Morecambe Bay." This gave him the opportunity to roam over a different countryside to the one with which he was familiar in the Kendal area. His maternal grandfather was a Methodist preacher, who was renowned for walking miles around the countryside to visit remote farms and hamlets, where he stopped to preach at different villages on his circuit. This religious input into Williams' family background, helped to redress the balance of the rest of his family tree, which William claimed included poachers and smugglers among his ancestors. The effects of the principles of the Methodist influence of his mother and grandparents, also ensured that he avoided alcohol through out his life.

Summer holidays also gave William the opportunity to attend sheep dog trials and sports meetings. The major meeting was always Grasmere Sports, which is traditionally held on the third Thursday in August. Summer holidays were shorter in William's schooldays, and in some years, the Sports often coincided with a school day. William was only aged ten when he walked to Grasmere for his first visit to the famous meeting. This was a practice he continued for the rest of his schooldays. The Sports were magic to William Palmer; the excitement, the good humour, the spectacle, and even the rousing music from the band that heralded the winners of the Guide races, made him determined to try the sport of fell racing for himself, when he was older.

William grew up in a tight knit community, where neighbours were regarded almost as extensions of his own family. Inevitably, there were feuds between some. In cases, these were so bitter that the cause of the feud was often lost with the passage of time, and a reconciliation only made on a death bed. But generally there was great friendliness in a community where doors were never locked, and there was always someone on a hand to help in times of trouble or difficulty. The youngsters were expected to run messages, not only for their own family, but for any of the older folk in the village. One of William's regular tasks was to fetch tobacco for some of the village men, from Kendal, which was over three miles away.

"As I was supposed [truly] to have no taste for tobacco, I was frequently sent for the stuff; on my return a grim old waller produced his steel rule, and measured the ounce purchased. 'Nine inches just-and that other young beggar browt me a short eight yesterday.'"

From an early age, William was happy to explore the countryside, wherever he happened to be. Often he wandered far from home to investigate hedgerows or woodlands.

His natural curiosity, and observation enabled him to become familiar with birds and wild animals of the countryside at a very early age. The pursuit of his interest in those days would have been frowned on in today's conservation and protectionist minded society, for he trapped song birds, he collected as many different wild birds' eggs as possible, and he pursued wild creatures in the name of sport. In those Victorian times, before the advent of an acceptance of nature conservation, it must be remembered that these activities were tolerated as part of the harmless occupations of the village lads.

He had a great admiration for one of the village men, Thomas Bowness Wright, whom he later described as "a wonderful observer and a student of nature. I knew him slightly when his beard was golden brown, and my chin had no hair at all." It was from men such as these that the village lads learned their country lore and the ways of its wildlife.

Among his youthful contemporaries, William had a great reputation as a tree climber. His skill was in great demand by other lads, to help them when they were searching out birds' nests or squirrels' dreys. William didn't flinch from tackling the unpleasant and uncomfortable climb up a rough, sticky pine or spruce trees, with their penetrating and irritating needles. He continued with his self appointed task, even though "it was not likeable." His simple, precautionary measures to reduce personal discomfort, were to tie bootlaces round cuffs or collar, to prevent the intrusive needles from working their way into irritating positions, but this was not always successful.

Sometimes, the spoils of his expeditions were not well received by his mother when he returned from his wanderings. On one occasion, he found a bat, hanging upside down in an old derelict building in a wood. He made a bed of moss for it in his satchel, and took it home. Unfortunately for William, but luckily for the bat, the creature escaped from the confines of the bag and gained its freedom in his mother's kitchen.

His mother "screamed, and there was some tumult" before the creature was chased out of the house. In those days, country folk believed that a bat's bite was poisonous, and a scratch from the creature was sufficient to bring on "fever or ague." The bat escaped unharmed; William was not quite so lucky.

Bonfire Night, November the 5th was the occasion for a celebratory half holiday from school when the afternoon was given over to the collection of material for the bonfire. Every child of school age was expected to take part in this activity and while the youngest foraged close to home, the older lads roamed two or three miles distant in search of fallen trees which were dragged back to the village. They ran the risk of battles with lads from other villages if they happened to encroach on their territory. The spoils gained in this way however were greatly prized, for often they had been bravely, if bloodily won. The bonfire was set with shavings from the carpenter's shop and the carpenter himself supervised the building of the bonfire with the wood that had been collected that afternoon. Any large timber was set aside and saved until later, when the bonfire had a good hold. When the time was ready for lighting and all the villagers circled round, the bonfire was lit by a candle from someone's lantern. Cheers greeted the rising flames, as sparks clouded from spluttering branches. The smoke swirled round the gathered crowd, reddening itching eyes with salty tears, as the flames leapt and danced. As the evening wore on, the blackened wood turned to soft grey ash. From the red heart of the fire, fingers of flame flickered upwards, passing shadows over ruddy faces. Small groups and couples, clustered in the intimacy of the darkened night, enjoyed the opportunity for a light hearted break in their routine way of life. Many of the youngsters did not stay to see the end of the bonfire, for sleep intervened; but most of the the older village folk saw the fire out to it's last flame. "T' bonfire cus but once a year." was their experience, and in the calendar of entertainments, the Bonfire ranked highly as a village event.

Fireworks as we know them today were not in existence then, but William and his brothers attempted to enliven one Bonfire Night by producing some of their own. Their home made brand of fireworks were concocted by following instructions that were given in a boys' magazine. Iron filings, brimstone, saltpetre and charcoal [most of these obtained from the village blacksmith], were mixed and divided into a number of tubes. The lads had not had the time to experiment with the quality or quantity of the charge they packed in the tubes, so for safety's sake, the "squibs" were set off some distance from the crowd. The resulting explosion on the beck side scattered sparks, pebbles and the remaining crowd, over some distance. No record is available as to what happened to William.

Christmas in William's childhood days had not the commercial overtones that it carries today. The schoolmaster recorded, "The Christmas Tree promised to the children definitely postponed until after the Christmas holidays," while the acquisition of a fowl for a family's Christmas dinner, was often left to chance.

Someone in the village took it upon himself to organise a raffle of geese or turkeys, and it was amazing how often the person organising the raffle ended up actually winning a prize. It was common practise to use numbered metal tokens that were made by the blacksmith for the raffles. These were available from year to year, unlike the disposable paper tickets that are used today. It was customary for the son of the organiser of the raffle to draw the winning token from a bag. On one occasion when the draw was not proceeding swiftly enough to satisfy the eagerly awaiting crowd, the child was urged to hurry up. "I can't find t'hot un" was his response.

In William's village, the raffle was decided by a system of three throws of three dice. The cobbler's shop was the traditional venue for this communal event, unless the crowd happened to be too large, and then they over spilled into the blacksmith's shop next door. Although some mothers frowned on the activity, nevertheless the event was considered of such importance that youngsters were allowed to watch the proceedings, that took place in the tense atmosphere of expectation. Sometimes they were even allowed to shake the dice for any absentee gambler.

On the first occasion that William was allowed to throw, he "scored the record for lowness." His father had already made his three throws, and achieved what appeared to be a winning score and was looking forward to taking home the prize of a fine bird for the Palmer Christmas dinner. William was given another chance to throw the dice for a distant, but absent gambler. This time, to a hushed room, and his father's dismay, William threw the highest score. The village blacksmith probably put his father's thoughts into words, "By gum Jim, but I'd skelp that young whelp if he was mine. He'll do hissel' out o' that Christmas dinner."

The skelping proved unnecessary, for on yet another throw, William also won second prize for the same absent farmer, who made sure that the Palmers received that bird for their Christmas dinner.

William's mother, who came from a strongly Methodist background, was against most sorts of gambling. Cards were frowned on, but the gambling of a shilling for a Christmas bird was one that she agreed to, albeit somewhat grudgingly.

William's mother was quite a religious woman, and although he was baptised in the village church of St. Oswalds and accepted into the Church of England, it was his mother's Methodist faith, and attendance at Chapel that he followed as a child. The kindliness, diligence and honesty of the Chapel folk made a lasting impression on him, as did the visiting preacher who held services in farm kitchens during the winter. In later life, William recalled memories of "a calm and confident religion, which graced the table where we worked and where also we took our daily bread."

In common with most families growing up in the late 19th century, access to medical care was not always available. Lack of knowledge, and suitable treatment meant that there were illnesses and accidents that frequently led to an early death in childhood. William's family did not escape, for his sister May died while only a young girl and a further tragedy hit the family in a different way when his father was involved in a severe accident at work.

The paper mill at Burneside offered regular employment for most of the village men. James Palmer gave up his full time farm work to take on a job at the mill, as it promised more regular hours and better pay. The accident to William's father occurred while he was working on a boiler, around which was some complicated cog driven machinery. In those pre safety conscious days, there were no guards to protect workers from injury caused by moving machinery. James slipped from his stance, and his leg was caught and trapped in the cogs. The dreadful injuries that resulted reduced his leg to a mess of crushed bone and torn flesh; there was a loss of blood and severe shock. After such an accident, and without prompt medical treatment, he was fortunate to survive with his life. His leg was so badly injured that it had to be amputated, and he stumped and hobbled on a wooden peg for the rest of his life.

Chapter 2 - THE MAN.

William's life when he left school followed the same course as that of most of the village lads that grew up in a rural community ; he became a farm worker.

Although physically he was not of any great height, he was sturdy in build. William was a strong lad, he was willing and good natured, and these qualities, allied to his ability to walk great distances over rough fell land, gave him the ideal temperament for working with sheep. He was patient yet persistent; quiet, yet determined, and he was not easily ruffled by the unexpected.

During the years that he worked as a shepherd, he gained the greatest enjoyment from the occasions that involved making the long walks to far distant grazing heafs. There, he joined with the other shepherds to gather the flocks to be driven back to their winter quarters in the valleys. Sometimes, when winter came early to the Lakes, sheep and men were taken unawares if the weather turned suddenly harsh. If this happened, there were occasions when even the sheep, hardiest of all the farm animals, were caught out by blizzards and overwhelmed by driving snow. It was at those times that William shared the responsibility with other shepherds, of seeking and digging out the buried animals; tasks which were hard, cold and sometimes dangerous.

"Day after day we now went on to the fells. On the ninth day, my comrades and myself drove a tunnel into a big drift piled against the craggiest part of the fellside. We took the drift lengthwise, and gradually sinking deeper, came to the level on which the sheep were. At the outset our proceedings were much hampered by the caving in of the walls and roof of the tunnel, and on one occasion we had to combine with those in the open to dig a way out again. By this method we came within reach of two score sheep."

When Spring came round again, to spread life and light over new emerging growth, he was quick to appreciate the freshness of a slow passage through hedgerows brightened with soft tender leaves, as the flocks, increased in number by the new crop of lambs, were returned to their native fells.

But the lad who continued to read Wordsworth, Shakespeare, Coleridge and de Quincey, as well as books on travel and mountaineering, was not content to stay a shepherd. He was excited by the beauty of the countryside and its many different features; he heard the music of the wind in wild and lonely places; it stirred his imagination and swelled his pride in his Lakeland heritage. Such emotions were not to be shared with sheep alone.

An opportunity arose for him to start work in the printing sheds of one of the Kendal based newspapers. There is no record as to which of the three publications that existed at the time employed the young Palmer, but it was probably the Westmorland Gazette, for his detailed account of a return to the printing sheds of his youth as recorded in his book, Byeways in Lakeland , identifies some of the still recognisable features of today.

The Forgotten Man of Lakeland

William Palmer's first job in the newspaper world was to help to set up the wooden blocks of leaded type which gave the ink stained accounts of local happenings. He became absorbed in the hustle and bustle of a busy work place. The frantic efforts of time - stretched journalists to meet a deadline, the smell of paper and ink, were all aspects of a life far removed from the slow steady plod of sheep over the freshness of rain soaked grass.

William Palmer found that his work among the press men was lively and stimulating. Each week brought fresh demands and developments. There was a steady, heady crescendo of activity among the workers that built to a climax as each edition went to press. Then followed a brief respite, which allowed a slow evaporation of tension before the excitement began all over again.

To a country lad, this was heady stuff; it was challenging and demanding. His involvement on the fringes of the news stories and features that others wrote, combined with his own enthusiasm for the printed word, fired his ambition to become a writer.

The beginnings were inauspicious, for his first post was as a junior reporter. This job was nothing grand, but it did set his foot on the first step of the ladder, of what was ultimately to become, a successful career in journalism. As a junior, he was given the coverage of some of the most boring and routine meetings that were held in the outlying districts of Kendal. He was sent to report on the proceedings of local councils, sports organisations and church activities that formed the humdrum assignments that fell his way. They may have seemed boring to one of a lively mind, but he realised that press coverage of such meetings was essential to enable local communities to keep in touch with what was going on and tackled his work with diligence, if not outright enthusiasm. Most of these meetings were dull in the extreme, but occasionally, sparks of controversy livened the proceedings, as happened on one occasion at Kirkby Stephen, "when a most comical meeting" about the installation of an electricity supply, ended up in a "free for all."

In his early writing days, the copy that William Palmer produced for editorial scrutiny didn't carry the excitement of a news story, or the in depth reporting of matters of local or national importance, but the demands imposed by his early years as a junior did give him the work discipline that was to be so essential when he became a freelance journalist with the Northern Syndicate, a Kendal based news agency.

As a young bachelor, who was already embarked on his chosen career, William Palmer was determined to spend as much of his spare time in pursuit of his hobbies and interests as he could. Long summer evenings or weekends at any time of the year, were spent walking, cycling or climbing in the Lake District, or venturing further afield into other parts of northern England. Often he was alone, but at other times was happy with the company of two or three like minded and physically compatible companions. Mostly these were professional men; among the company were schoolmasters, lecturers, or those connected with commerce or business. Their conversations were brisk and lively; ideas sparked between them, and lit the base of the fire for some of William Palmer's freelance writings.

One of his friends was the Cumbrian writer, and former editor of the "Penrith Observer" newspaper, Daniel Scott, whose book "Bygone Cumberland and Westmorland", which was published in 1898, has become something of a standard work of reference on the old customs of the two counties. Daniel Scott had an extensive knowledge of the Pennine area and was generous in sharing his expertise with the youthful Palmer. "I accompanied him on one or two walks through little known moors and passes, and he was full of the most wonderful stories," wrote William Palmer

Among his circle of friends were other walkers and climbers who also lived in Kendal. Together, in 1906, they joined a newly formed mountaineering club and became original members of the Fell and Rock Club Climbing Club of the English Lake District. They joined in the regular monthly meets which became part of the Club's annual programme that were held in different parts of the Lake District. It was not unusual to have up to thirty members attending these meets, which was a large gathering in those early days of limited transport.

William Palmer was not one to join in these gatherings on a regular basis, for although he attended the Coniston meets whenever he could, he preferred his ventures further afield to be taken in the more intimate companionship of a small group. On other occasions, the companions that he acquired were "heterogeneous comrades picked up on the way." These may have been other climbers or walkers who happened to be sharing the lodging accommodation that he used when he embarked on long walking expeditions. Such casual encounters between like minded characters, gave, for short periods of a few days, the reassurance of intimate friendship; however brief.

There were other encounters of a more dubious nature. When the annual horse fair was in full swing on Brough Hill, he often spent a night bedded down among the horses of the travelling folk. "It was a tough experience, certainly, but the encampment was the weirdest, unlike anything I know. One wet night it was hard to mind one's business. The first time I saw a Gypsy belting his lover I was pretty sore and wanted to get at him." But he was warned not to interfere. He was advised by a wiser, and more experienced companion that the brutal activity was a ploy to persuade someone with a purse to get into a fight so that the purse could be removed by the girl, "while he cracks your head."

"Mostly we walked elsewhere than among the campers, and spent our time grooming and tending the horses that had come down from the hills."

William Palmer kept himself fit for his outdoor activities by walking to and from work. When he lived at Bowston, this involved a distance of about four miles at each end of the day. Sometimes, he lengthened his journey by taking diversions on the way home, especially when he was building up his stamina for more demanding expeditions.

William Palmer gave up his bachelor lifestyle at the age of 24 when he left Bowston to move into Kendal, where he set up home at Lake Road Terrace with his wife, Annie Ion. Annie was two years older than William and although she was a member of a Kendal family that had business connections in the grocery and shoe trades, Annie worked as "a fitter in a boot factory", which is still producing the high quality footwear for which Kendal has become noted, and is known as K Shoes.

In more than one of his books, William disclosed some of the locally held secrets that went to make for a successful marriage. One of these was to wed a girl from one's own locality, another was to apply the "toffee test." The latter was based on an old tradition of "toffee joining" whereby a few lads and lasses, who were in many cases hired to local farms, got together to make toffee. These were very often spontaneous affairs; little more than an excuse for a night out, when the young people of farm or village, gathered together after the day's work was done. The "toffee joining" was regarded as an impromptu social evening. Word of mouth spread the news quickly throughout the neighbourhood when a toffee joining was planned. Those taking part, were expected to take it upon themselves to provide all the necessary ingredients of butter, cream, sugar, and treacle, needed to make the toffee.

At the appointed time, all the young folk appeared at the designated meeting house. The quality and texture of the toffee was heavily dependent on what ingredients happened to turn up, and in what sort of quantity. They all took a turn at the "stirring" and had a "lick o' the spoon." These were occasions, in a restricted social calendar, that gave a lad and lass the rare opportunity to pair off, albeit to pull and stretch the toffee, but many a sly kiss was taken on pretence of licking the spoon.

"Aye toffee joins - them's the spooart an' it's a rare time to watch the lasses. Some on them git that clarted up that it isn't safe to touch their apron string else thou'll git stuck. Choose yan on 'em as wets her fingers afore she dabs in toffee, an thou'll be reet" was the advice that William was given by one old character.

William Palmer didn't choose his wife by this test, although he was assured by many an old countryman that it was the way that their serious courting had begun. William married a lass from Kendal, which was near enough to Bowston to be classed as his own area.

Like William, Annie Ion came from a family of five children and she shared many of her husband's own interests. She was an enthusiastic botanist who loved exploring the countryside. She enjoyed cycling and walking, and shared William's love of the Lake District. She didn't have the enthusiasm or skill for rock climbing that her husband displayed, although she did eventually demonstrate that she shared his interest by becoming a member of the Fell and Rock Climbing Club. Annie Palmer preferred to take her enjoyment of the mountains in the more leisurely and less demanding activity of walking.

The wedding of William and Annie took place at, what was then, the Zion Church in Kendal on the 30th September 1901 "according to the rites and ceremonies of the Congregationalists." The building still stands in Kendal today, but Zionism has been replaced in the town by the United Reform Church movement. A comment by a modern day registrar is that "Half of Kendal must have been at the wedding" as the marriage certificate is signed by no less than seven witnesses.

Annie brought good "connections" to William Palmer as her dowry, for her family were well respected within the town and many members were successful in trade and business. Most were involved with cobbling and shoemaking, while others were in the grocery trade. Annie's grandfather was a former manager of Stott Park Mill, one of the numerous bobbin mills that were to be found in the coppice woodland area of the southern part of the Lake District. On the more tragic side, her youngest brother, William Wharton Ion, volunteered to serve as an ambulance man in the Boer War and was only 18 years old when he died in South Africa.

Annie Palmer was a quiet woman, and of a very dependable nature, who accepted the countrywoman's practice of keeping well within her husband's shadow. But if she was unobtrusive, she was always there to support him in any venture on which he embarked. She was ready to listen to him when times were difficult, and she was there to encourage him in his work.

Annie was artistically gifted. She played the piano well and was skilled at all sorts of craft work. She created leather work items and gemstone jewellery, and was able to make small items of furniture such as stools with woven rush seats. Her skill as a needle woman ensured that she able to make and mend clothes for her family with neatness and efficiency.

She was also a knowledgeable woman in her own right in country matters, and was able to add some of her personal contributions to William's books, especially those sections that included references to the flora of the countryside. She complemented her husband well, for she was the perfect steadying influence to his more ebullient character. She was prepared to back him up in anything that he did. William appreciated her support, and frequently made reference to her in his books as, "the lady at my elbow."

Their marriage proved to be a happy one for the fifty or so years that it lasted. William and Annie had two daughters, the eldest, who was born in 1905, was also called Annie, [a traditional name in both the Palmer and Ion families]. Their second daughter Jean, was born seven years later in 1912. They became a very close family unit, sharing many interests and activities together.

William and Annie not only introduced both their daughters to physical activities related with the countryside, but also encouraged both girls to develop their musical and artistic talents. They tried to ensure that their daughters gained the maximum enjoyment and fulfilment from their lives in those early carefree days. William and Annie ensured that every year, time was made available to spend family holidays walking or cycling, and camping in different parts of Britain. Very often, however, these holidays assumed a working nature for William, as he took the opportunity to acquire and assimilate material for his articles and books. Unbeknown to them all however, in those early carefree days, William Palmer had already bequeathed to both his daughters a devastating inheritance of an incurable disease, which manifested itself much later in the lives of William and both his daughters.

In the second decade of the 20th century, William Palmer had begun his lifelong interest with the Scottish mountains. He made regular visits there for climbing and walking during the winter months; on some occasions he took his wife and young family with him. It was not until several years later, when his daughters were grown up, that,

The Palmer Family (L to R)
Daughters Jean & Annie, Annie (Senior) and William

William Palmer continued to remain
a countryman at heart.

on the insistence of his wife, they were taken to see the beauty of these mountainous areas in the summer months; all their previous visits having been made at chillier times of the year.

As an indication of Annie's own enthusiasm and capability for mountain walking, she devised her plan to ascend the highest peaks in all four countries of the British Isles in one year. Her programme for the family was to climb Snowdon at Easter, Scafell Pike at Whitsun, with Ben Nevis and Slieve Donard, [Ulster], at the beginning and end of August.

With such enthusiastic parents, it was inevitable that a third member of the Palmer family eventually became a member of the Fell and Rock Club, for daughter Annie joined her father as a life member of the Club, although she was never very much involved with their activities.

Both William and Annie were mindful of the limited levels of formal education that they had received in their village schools, and they ensured that their daughters had the opportunity to follow their education through to the highest level. Both girls carried on their chosen studies at university level, which was uncommon practice for girls during the 1920's and 30's. Annie went on to successfully complete her years of study, and gained a good science degree from Liverpool University which led to a post as a science teacher in a Liverpool school. However, when Jean followed her sister's path to university, she found the stress of an academic course too demanding and opted to change course for the more practical training of a social and youth worker, which led to a successful career.

William and Annie, brought up their daughters in an atmosphere of loving care, supported by their own, firmly held moral principles. The strong discipline of Williams's own upbringing, and the Methodist convictions implanted by his mother, stayed with him throughout his life. It was only on extremely rare occasions that he drank alcohol, even though he frequently took advantage of the hospitality of many an inn. He never smoked and as a lad, he was in great demand to run errands for the village men, to collect their weekly tobacco or snuff from Kendal. They knew it was safe with him.

The cornerstones of religious principles, formed an important part of William Palmer's life. Although he was baptised in the Anglican faith, his early life was strongly influenced by Methodist teachings. After he and Annie were married in her own Xion church, his wife turned to her husband's Anglican faith, but they both later found great tranquillity and comfort through attending meetings of the Society of Friends. Although they adhered for many years to the principles of the Society, they never did become Quakers, and later they drifted away from that persuasion to return to the Anglican faith. A relative commented, "In later life, they became very high church. It was almost as though there was a constant spiritual searching throughout his life."

William Palmer continued to remain a countryman at heart; it was an ethos he maintained in his manner and speech. He was a life member of the Cumberland and Westmorland Dialect Society. He was able to converse easily with farmers and shepherds in his native Westmorland dialect. His easy familiarity with country men immediately made him accepted at gatherings or meets, where he joined in their conversations with a knowledgeable authenticity. He was no outsider to these men; he was accepted as one of their own.

Yet, as his aspirations increased, he circulated on a higher social plane in both work and leisure activities. His friendships extended to include those with professional, business and academic members of the establishment. Both mentally and physically he left his village background behind him, in the course of professional enhancement, although it was never far from his mind, and he never, ever, forgot it.

Among his new found friends and acquaintances, he discarded his native colloquial turn of phrase, yet he was able to return to it, whenever the company and occasion suited. Throughout his life, he projected a countryman's image by his rough tweed mode of dress. He was hardly ever seen in anything other than his comfortable breeches or plus fours. This countryman's garb only gave way to wearing business like suits when his work took him to London, during, and after, the second world war and then, he dressed as a city man.

An earlier war, that began in 1914 left its scars on William Palmer. When the war started, he was aged 37 and was too old to be eligible for call up to military service. In spite of this, he volunteered "for the colours" on a number of occasions, but each time was bitterly disappointed to be rejected because of defective eyesight.

In a wave of almost hysterical patriotism that swept through country and county, there were recruiting crusades that were given wide publicity in the local press. At the beginning of November in 1914, there was even a recruiting sermon preached in the Zion Church in Kendal. William despaired as the war continued, and the fighting took its toll of the regular army, demanding thousands more to replace those killed.

The response from the men of Westmorland and North Lancashire to come forward to form the new army in 1916, was such that the Kendal based recruiting officer, Captain Long, was overwhelmed by the numbers that came to the depot. He appealed for well qualified clerical workers to assist as recruiting officers. William Palmer, who had been rejected yet again to join "the colours", was one who responded to Captain Long's appeal, and he was appointed to help out at the Kendal depot.

It was somewhat unusual to employ a civilian as a recruiting officer, for most of the men appointed were of considerable military service. They were usually in the 40 - 50 age range, with a minimum ranking of a non commissioned officer. They were men who could display authority and command respect, but the demand for replacements to the army was so great in 1916, that any experienced former soldiers that could be withdrawn from non essential duties were transferred, and deployed to train the mass of new recruits.

"In a small town like Kendal, the recruiting office was likely to be swamped by a rush of men wanting to enlist, so it is possible that a local man with good administrative skills could be appointed." [Curator King's Regiment Museum, Carlisle Castle].

The recruiting system for the first world war was closely connected with the established Territorial Army. For many men, it was a simple matter to transfer from the

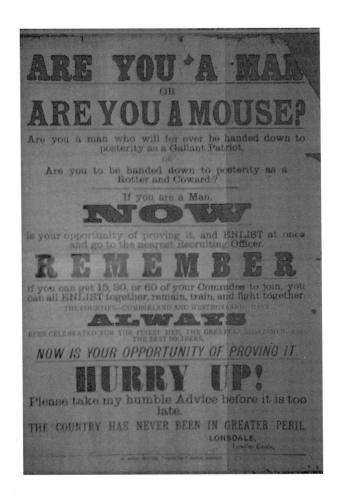

First World War Recruiting Poster

role of being a part time soldier, to becoming a full time serviceman. The Territorial Army depot in Kendal was affiliated to the Border Regiment, and two battalions of that regiment were raised in the town, the 1/4th and the 2/4th. Both served in India where they saw action on the North West Frontier. Other recruits were sent as replacements to the many different regiments or corps fighting in Europe, as the devastation continued to take its toll. The Kendal depot also signed up recruits from the industrial areas of North Lancashire, where there was a surplus of men for that county's own home based regiments.

It was a difficult time for everyone. Food was scarce and rationing had been imposed. There was hardly a family in the land that did not suffer from the strain and tension of being broken apart, which in some cases was forever. News of troop movements, and the results of fighting were slow to filter through. The war years imposed a period of worry and sorrow; austerity and hardship; frustration, anger and loss.

William Palmer carried out his duties as a recruiting officer alongside his own work as a journalist, writer and his editorship of the Fell and Rock Journal. Something had to suffer and it was in his editorial notes in the war issues of the Journal, that gave the indication of the pressure to which he was subjected. Inevitably, the tension increased, as the wearing effect took its toll. It was like the incessant weight of a fallen climber's body on a straining rope that chafed against an unyielding rock, before the final break. His writing suffered. The apologies he made for the delayed issues of the Journal were frequent; yet, it was an achievement and testimony to his determination and tenacity, that the Journals continued to be published throughout the war years . The demands of work in other fields left him with little time and enthusiasm to devote to his own writing, for only one book was published between 1913 and 1926. These thirteen years spanned one of the leanest periods of his writing career.

The effects of the war years had a traumatic effect on William Palmer. Although those who remember him, recall that he was not a man to show a great deal of emotion, it is recorded that he broke down completely while writing the editorial notes for the Fell and Rock Club Journal of 1916. This was the year of the battle of the Somme, when thousands of lives were lost. Pages of photographs of young Kendal men who had been killed in action stared back at him every week from the Westmorland Gazette. William Palmer was so emotionally disturbed by the resulting devastation of the war years, that he could not complete his editorial piece for the issue for the year of 1916. It came at the culmination of a period that had been beset by difficulties. By way of explanation for the delay, he submitted an Editor's Apology for the lateness in the publication of the Journal.

"Our Club has passed through months of fierce trial. Some members have passed, all too soon, into the Great Beyond. Many are standing in hourly jeopardy. Duty is being nobly done. The destiny of our Nation is sure so long as its strong men dare the great sacrifice for the sake of it's honour. Yet, in this hour of gloom and pain, one cannot but think again and again of the Eternal fells, of the great sympathies we have found there, - and one looks forward even to the great day when, with Victory, the remnant shall meet again in the shadow of the mighty rocks." [F&R Journal 1916].

At this point; in the writing of the Editorial Notes, his wife, on whose strength of character he relied so much, had to finish the piece for him.

She wrote, "At this point my husband has completely broken down. The editing of this Journal, added to his Recruiting Office duties, has been a great strain. The memories of brave climbers and the apparent certainty of a long farewell (may it prove only to be a short one) has caused him to lay his pen aside in tears. He has spared neither time nor consideration, and his heart has been thoroughly in the work. No one feels more than he the tremendous loss of great and good comrades, and he looks forward to the day, when his services may, after many rebuffs, be of direct help in protecting the honour of our own England - Annie Palmer."

In this emotional manner, was the Journal of 1916 completed for circulation among the members of the Fell and Rock Climbing Club. William felt that it was essential that the circulation of the annual report to the members was continued, for he regarded the Journal as one instrument of sanity in what was to many, an insane world. Although there were delays in completion, and the contents were not as full as wartime restrictions imposed, he ensured that every member of the Club who was on service in the armed forces, received his copy whenever possible.

He saw it as part of his editorial duty to keep records of the movements of those members who served abroad. He wrote them personal letters to keep in touch with life at home, and he urged them, whenever possible, to send in contributions to the Journal, however meagre they seemed at the time.

During the war years, which encompassed half of William Palmer's term of office as editor of the Journal, he had the unenviable task of either writing or editing the obituaries of former members of the Club. The waste and futility of war hit hard with the obituary of RH Whitely, 2nd Lt. 7th Royal Scots, who was an Ulverston member of the Fell and Rock Club, and was killed on 19/7/1916, "thus having his dearest wish fulfilled that he might serve at least three months."

Probably the most devastating news received by William Palmer was of the death of Lehmann J Oppenheimer who was aged 48 when he died, 8 years older than William Palmer. Oppenheimer was an unlikely man to go for a soldier, but he was conscious of his German sounding name. He wanted to prove his loyalty to his country in the most positive way that he knew, and that was to enlist. Oppenheimer was a sensitive man, a writer, poet and painter, who conceived that it was his duty to go to war. Once he made that decision, nothing could persuade him to do otherwise. The thought of doing what he considered to be unessential civilian work of designing churches, was an anathema to a man of duty and honour.

Oppenheimer's death hurt William Palmer greatly for he was a former climbing companion and a good friend with whom he had shared the pioneering and exploratory work on the difficult new north west route of Pillar Rock. With Oppenheimer's death, those former exhilarating days were gone forever as far as William Palmer was concerned, for he never seemed to regain his former enthusiasm for climbing, once the war years were over.

William Palmer had another upsetting role to fulfil in his position as editor of the Journal. This was the task of disposing of Oppenheimer's personal collection of climbing, and mountain related books. To those who love books, their ownership is not only that of a collection of printed words on a page. Many books assume deeper personal meanings; they become friends. They provide collections of memories, with an intangible hold on what has been. In disposing of Oppenheimer's books, it was a partial disposal of the man himself.

Did William Palmer torment himself by the fact that Oppenheimer had died, while he still lived? Oppenheimer's death occurred in a place far removed from the Lake District that he loved. He was the victim of a poisonous gas attack. When William Palmer heard the news of Oppenheimer's death, he still breathed the clean cold air of Lakeland, while his friend had choked and died. Another fact that may have weighed on William Palmer's mind was that although Oppenheimer was an older man, he had volunteered, been accepted and served, while William had volunteered, and been rejected.

Another casualty of the first world war was a boyhood hero of the young William Palmer. Frederick Courtenay Selous, was a former big game hunter, sportsman, adventurer, traveller and a typical "Boys Own Hero" character, whose tales of daring exploits enthralled the young Westmorland lad. "As a youth I was thrilled by the handling of some of his letters and papers; I heard him lecture; gave a loving reading to all his books I could find. Life never permitted me to copy his journeys and his feats of sports." Selous was shot in the head while serving in East Africa in January 1917 and yet another of those men that William Palmer so admired, was lost.

So, although William Palmer was not directly involved in the actual fighting of the first world war, he did not escape from the ordeal mentally and emotionally unscathed. As with others, the trauma left its mark. It is perhaps as a result of the effects that war years left on William Palmer, that although he lived through periods of two world wars, they are rarely mentioned in any of his books.

After the first world war ended in 1918, the Palmer family moved to Liverpool. These new surroundings presented him with greater opportunities in his field of work. Although his years with The Syndicate, in Kendal, enabled him to send stories and articles to a wider readership than his immediate surroundings, the Westmorland town was far removed from the hustle and bustle of national news. For a journalist with ambition, city life presented a greater challenge.

The move to Liverpool was in a way, a natural progression for William Palmer. It was a major city in industrialised Lancashire, yet not too far away from the families and friends in the Lake District. It had the further advantage of having the mountains of North Wales and the Peak District close at hand. Liverpool was an ideal centre for further explorations of these areas, and there was always the possibility of something newsworthy happening in a busy city.

The Palmer move to Liverpool gave the opportunity for new friendships to be forged, that helped to heal the wounds of memory of old friends lost. Yet William Palmer felt, as did many other members of the Fell and Rock Club, that it was vital that those old friends should not be forgotten. He supported the venture that was led by an officer of the club, H P Cain, that a memorial should be erected to honour those members of the Club that died in the war. As a result of an appeal to Club members to support this venture, sufficient money was raised to purchase an extensive area of fell land as their memorial. Through his writing, William Palmer took the opportunity to pay tribute to the generosity of Herbert Walker, a Seascale man, who made the memorial a possibility, by selling the land to the Club for the almost nominal sum of £500.

William Palmer and his family quickly adapted to their new life in Liverpool, although it was so different from their former home in Kendal. There were opportunities for the girls to develop their artistic and musical talents; there was more advanced schooling available that led to university careers, and there was plenty of freelance journalistic work to ensure that William Palmer was able to support his family in comfort. But in spite of the seeming advantages that city life brought to the Palmer family, it was many years before his book writing resumed its former and early momentum.

William Palmer quickly settled into his new post as a freelance writer with the Liverpool Daily Courier, which was later taken over by the Liverpool Daily Post. The editor at that time was Lewis Chesterton, a man in whom William Palmer found a kindred spirit, for they both shared many of the same outdoor interests. John Chesterton, the Westmorland born son of the newspaper editor explained, "While father was alive, he was able to put useful contacts in Palmer's way."

The Palmers made their home in Salisbury Road, Liverpool and John Chesterton, recalled going to visit them with his father, shortly after the family moved to the city. John was only about six years old at the time, but he could recall a clear impression of the Palmers and their home. He remembered that the house was one of quite a dingy looking terrace and the interior of the house was dark and gloomy, with solid, functional furniture. "My first impression was that the family must be quite poor" he said. That impression was also supported by the fact that he clearly remembers the gas mantles being lit in the Palmer living room. The wheeze of gas emerging from a swan necked pipe, a faint splutter, then a pop as soft yellow light hissed the room to life, scored his memory.

This however, was only a small boy's recollection and was based on a comparison with his own home circumstances. "We lived on the other side of the river (Mersey)," where electricity had already been installed in his house. "I thought everyone had electricity in those days. I thought it was normal. I didn't consider our family to be terribly well off."

Annie and William Palmer made quite an impression on the memory of a young boy, but for quite different reasons. Annie Palmer, whom he later came to know very well, struck him in those days as being a very quiet woman. He could not remember her ever saying a great deal, at any time. She was a woman who always stayed in the background; but she was a tower of strength to her husband and supported him in all his ventures.

John Chesterton remembered William Palmer as being quite different in character to his wife. "He was very rumbustious." John Chesterton recalled that "In physique,

William Palmer was a very burly figure of a man. He was broad shouldered and stocky. He gave the impression of having great strength." This image presented a very intimidating figure to a small boy on that first meeting, an impression which continued for some time, until John got to know William Palmer much better.

During the 1920's, although William Palmer continued to supply newspapers and magazines with sufficient contributions to earn a living, his own book production was sporadic, for only two small pocket type guide books were produced for the firm of A.C Black, together with the updating of one of his early successes, "The English Lakes" which was published by the same firm.

During that decade, however, one of the highlights of the Palmer family life was the marriage of his daughter Annie, to James Buchanan. James, who was known as Hamish to the family, to distinguish him from his father of the same name, and Annie were of much the same age. They had attended the same school, and they shared many interests that included cycling, camping and all sorts of activities connected with the general outdoor life. They had both studied science at Liverpool University, and while Annie used her degree in marine biology to become a teacher, until the time of her marriage, Hamish's profession was that of a chemist. They set up home on the opposite side of the Mersey, in Birkenhead, and much to the delight of William Palmer and his wife, their only grand child was born in 1932.

The years of the thirties, brought about some of William Palmer's most prolific writing, when he produced 14 new books within a ten year period. It was during this time that the Palmers moved house yet again. With both daughters grown up and leading their own lives, William and Annie left Liverpool to return to the Lake District in 1934. They made their home in a typical Lake District cottage at High Wray, near Ambleside. It seemed as though the well loved and familiar surroundings of his native fells, lakes and valleys gave further impetus to his writing career, for in the five and a half years that he lived at High Wray, William Palmer produced at least seven books. One of these "The Verge of Scotland" was dedicated to his own "Buchanan Clan", which comprised his daughter Annie, his son in law Hamish, and granddaughter Christine.

During the years that they lived at High Wray, the Palmers were often visited by their "Buchanan Clan." Some of those memories of her early years have survived with his grand daughter, for over fifty years later, when Christine was asked to recall the most vivid memory about her grandfather, it needed very little mental searching to respond with "his bristly chin." She added "He was a warm, cuddly person, who used to give me many a hug and a scrub" [a family reference to a rub from his bristly chin]. William Palmer , in common with many country men of his generation, wasn't too particular about presenting a clean shaven appearance to the world at large. A weekly scrape with a cut throat razor, taken at his favourite barber's shop in Kendal was, in his opinion, enough to keep the worst of his whiskers at bay.

Christine recalled that "He and Grandma Palmer were wonderful people who used to take me out for walks and always find something interesting to show me." One memory that stood out vividly in the mind of a young child, was that on the occasion of one of her visits to High Wray, she was taken to a bluebell wood. "I can remember it to this day. We went a little way into the wood where there were a lot of beech trees on a slope. The sun was shining through the leaves. The flowers were all bluing away like mad."

With the onset of the second world war, William Palmer yet again attempted to find some way in which his services would be of use to his country, as he had in the 1914 - 18 period. One of his friends and contemporaries was fellow writer Sir Hugh Walpole, who wrote to William Palmer.

"You've been a jolly good friend to me for a long time, and I realise it." When the possibility emerged that Sir Hugh Walpole may be offered some work at a Ministry, William Palmer contacted his friend in the hope that he might be able to offer some way in which William could help the war effort.

He wrote to Sir Hugh on 2nd Decemberr 1939

Dear Sir Hugh,

I am awaiting information about which Ministry you have started, and with which I am to start!!!!

Freelance work has stopped, and I am trying to get another job. The local powers do not seem at all keen to provide employment, despite what I have done for them.

If you can introduce my name anywhere, please do so. I expect to be at the YHA meeting at Grasmere tomorrow Sunday, to keep in touch with the youngsters.

The old Lakelanders are beyond hope."

William's disillusionment turned to disappointment when he received a reply from Sir Hugh Walpole three days later.

"I'm afraid it's hopeless just now. After months of flirting with the Min. of In. it has all come to nothing. I've just got a big job as chairman of the Books and Manuscripts at the auction at Christies for the Red Cross but that would be of no use to you I fear.

If I hear of anything I will certainly let you know.

Yours sincerely,

Hugh Walpole."

William Palmer wrote to other eminent people in the hope that they would listen. "I am fit and well enough to give a good deal of help in the wilder parts of England, and my desire was to get in touch with some authority able to use my services. I may be wanted later on, and I am here."

He gained no response other than to be advised at the age of 62, "you would be wise to carry on in your occupation and give such time as you can spare locally." This was from

Norman Birkett, the great barrister and benefactor of the Lake District, who later became known as Lord Birkett.

When the second world war broke out in 1939, members of William Palmer's family, like many others, were scattered throughout the country. His youngest daughter Jean, who had trained to become a social worker, was in London. She later joined the ATS and one of her duties included teaching the new recruits, the rudiments of car mechanics. The Buchanan Clan were in Liverpool where Hamish had added the job of firefighter to that of a chemist, while William and Annie had moved from Ambleside to live in a large house in Kendal.

The importance of Liverpool docklands and shipyards ensured that they were a major target of wartime bombing raids, and Hamish sent his wife and daughter to live in the relative safety of the Lake District, where they were well away from the nightly visitations of the Heinkels and Junkers. They took a house in the Kendal Green area, but eventually moved in with the Palmers, who had moved to live in a large terraced house that overlooks the town, 7 Cliff Terrace.

There they stayed for most of the remaining period of the war years. They shared the large house with William and Annie, until William and his wife moved south to London in 1943. Their stay in the capital city was a short one, and it is almost certain that the bomb damage caused by the blitz was the reason that William Palmer and his wife only made 5 Earlesfield Road, Wandsworth, a temporary home, for they returned to the Liverpool area in 1944. This time it was to stay on the other side of the Mersey at 46 Grosvenor Street Wallasey. Once again, their stay was of a short duration for after the end of the war, in 1946, the Palmers were back in London at their Earlesfield address.

Although over the years William Palmer had shown that he was a man capable of turning his hand, or foot to many different activities, there was no particular field in which his star shone brighter than all the others of his time. Even his long distance walking feats, remarkable though they now seem, were not outstanding achievements among his contemporaries. Others equalled, and then surpassed some of his walking records. But although many could match his skill, ability and performance in one particular sphere, few could match the variety and range of his abilities. He was successful in many fields, and it was this ability to maintain a high level of success, that raised him to the stature of a remarkable man.

He combined the prolific output of a successful journalist and writer, with the practical hands on skills of a farm labourer. He matched the rock climbing of some of Lakeland's best, with the enjoyment of simple walks over rough fell. He was a nature lover, yet he pursued the fox and otter with determination and endeavour; he gained great satisfaction and excitement from being in at the kill. His involvement in badger digging would be held as intolerable by many of today's conservationists and those who press for the cessation of "bloodsports"; yet to William Palmer, there was nothing immoral about what he was doing. It formed an acceptable part of his countryman's background, it was an established part of the system in which he had grown up.

He was a man of contradictions, for although he was born of humble origins; he sought honours and recognition far beyond what his village family conceived to be possible. At home, he was gentle, yet in the work place, ruthless. In many respects, he was like the infamous character of Jekyll and Hyde, for the more information about his personality is uncovered, the greater the realisation of its complexity becomes.

As the years move towards the latter part of the twentieth century, there are few people in his home town of Kendal and further afield, who remember William Palmer to give their direct accounts and impressions of what he was like. Like most writers, he does reveal something of himself and his character through his work. Rarely, however, is it obvious and direct; but in slight asides and with veiled implications, he opens up his inner self to reveal the man.

Qualities of self reliance, determination, humour, kindness and gentleness emerge from the printed lines. The image that he adopted for the work place was one of self enhancement, as he pursued a constant search for success and achievement. The letters of status, that were appended to his name, increased as successive books were published. Yet, those appendages that seem of worth when regarded at face value, had a hollow quality about them. An investigation into how they were acquired in William Palmer's case, revealed that they denoted membership of societies, for which nothing more was required than proposers to support his application. Once that had been approved and accepted, all that was left was the regular payment of a subscription, for annual, or life membership. In this way, William Palmer acquired an impressive array of letters appended to his name; FRGS, MBOU, FSA [Scot].

Among William Palmer's circle of friends and colleagues, there were many, who were themselves members of august societies and organisations. Through their willingness to act as proposers, they provided him with a stepping stone to higher things. It seemed as though William Palmer was not merely content with being a success, he had to have tangible proof that he had become successful. In 1930, his sponsored application, and payment of an annual subscription brought about his Fellowship of the Royal Geographical Society, and gave him the first opportunity to append letters to his name. No longer was he plain William T Palmer, he became William T Palmer FRGS. Those four capital letters after his name presented a more prestigious image than had hitherto been the case. This was followed by his acceptance for membership of the British Ornithological Union in 1939, with the letters MBOU; and finally his Fellowship of the Society of Antiquaries [Scotland] was approved in 1943, which entitled him to use the letters FSA [Scot].

Through the circles in which William Palmer moved, especially in his Liverpool and London bases, he cultivated work and recreational connections, to provide himself with a readily available supply of suitable contacts for enhancing his career. There is no doubt that he made use of anyone who could help him in his professional capacity. Among those who formed the circles in which he moved, were many men, and women of learning, position or rank. Was there an inner feeling of inferiority by this son of a Westmorland shepherd?

DIED.....*Dec 1954*.....
RESIGNED..........
REMOVED..........

CERTIFICATE OF CANDIDATE FOR ELECTION

Name...... *William Thomas Palmer*

Description...... *author*

Residence...... *69 Salisbury road, Wavertree, Liverpool*

being desirous of becoming a FELLOW of the ROYAL
GEOGRAPHICAL SOCIETY, We the undersigned recommend
him as a suitable candidate for election, on the following grounds :

For nine years Editor of the Fell and Rock Journal;

Member of the Fell and Rock, Wayfarers, and other outdoor clubs;

Chairman Liverpool District Assocn, Campers Club of GB & I;

Writer on outdoor and Nature

Dated this *tenth* day of *January* 1930.

WILLIAM HOWARTH .... F.R.G.S. { *From personal knowledge*

S. BUTTON .... R.G.S.

Received *15 Jan. 1930*. Proposed 2 7 JAN. 1930 Elected 1 0 FEB. 1930

*Facsimilie of the Cerificate granting
William Palmer the right to append FRGS to his name.*

At that time, membership of some notable societies was not earned by demonstrating any remarkable scholarship or evidence of special study; neither was a specific contribution in the furtherance of knowledge required. Yet the appending of their impressive looking array of letters to the Palmer name created a false impression. With the benefit of hindsight, the acquisition of these "cosmetic" honours is a disappointment. It seemed in retrospect, hardly necessary for William Palmer to seek them out, for he made his mark in other ways.

Further investigation revealed that although Palmer maintained his membership of these societies throughout his lifetime, he made no specific contribution to furthering the knowledge or study in the fields of geography, ornithology, or antiquary , other than in his writing where, conscious of the popular appeal of their contents to his readers, he kept the information at a superficial level.

John Chesterton recalled that while William Palmer had a very genial nature, and was kind and helpful to young people, he was very assertive where his profession life was concerned. John conceded that this was an essential quality for success in William Palmer's chosen career, for in his opinion, William Palmer was very much a man of ambition, and he had to sell himself, to get noticed. And yet John Chesterton's impressions of William Palmer as a private individual preclude the fact that any personal vanity was involved in the acquisition of such letters. In his opinion there was no conceit in the man, but he was determined to make a success of his career and would use any means to ensure that success.

Did he feel that some tangible sign of success was essential to his image? William Palmer seemed to be forever striving to achieve something better; as if he had something to prove to the world at large.

But the private man presented a different character; where no proof of his worth and ability was needed to family and friends. He was a quiet man, caring and compassionate to those around him. He didn't boast about his achievements at home , or make a fuss about what he had done. Even today, surviving members of his family know little about the full extent of his writing. Their existing knowledge is based on the possession of one or two personally gifted books that have been handed down through their intervening generations. "It's a Palmer trait that we don't boast about things. Palmers can't do with a lot of fuss," explained the late Jack Palmer who was Wiliam's eighty year old nephew, when I talked to him in his home village of Burneside.

As with most people, the passing years brought changes to many of William Palmer's attitudes. Mellowed by age, his former enthusiasm for "country sports" declined. Compassion was extended to some of the animals he had pursued with such relentless determination in his earlier and younger days. He changed the direction of his interests from hunting to studying. Over the years, he had built up a wealth of knowledge about the countryside and he widened his range by sharing his wife's interest in botany, and extending his own understanding of geology. He accumulated reference books to broaden his knowledge and was always willing to pass on this information to younger people to help them to learn the skills of outdoor pursuits. He realised how important it was to nurture their fledging enthusiasms, so that their interests continued to thrive. In this respect, he was only passing on the sort of help that he had received from older men in his home village, when he was a lad.

Throughout his writing career, William Palmer wrote of "the good tidings of health, majesty, beauty and friendliness in the open air, by dale and moor, ridge and height. So have others who have more skill and work to better effect. Artists and photographers as well as writers have shown the glory and colour, the light and shadow, passing clouds and dappled showers. There would be disgrace indeed in failure if the heights and dales and moors remained unvisited and unknown."

But these particular words were written in "Bye Ways in Lakeland" in 1952 in response to criticism that William Palmer, among others, had encouraged people to visit parts of the country that showed great beauty, and interest. They were accused of causing overcrowding of the hills of Britain. Over 40 years later, the same level of criticism is being levelled at producers of guide books for encouraging people to take to the high ground. They would do well to read other words of William Palmer, who never forgot the joys and delights he found in the open countryside, and wanted to share his experiences with others.

"The hills belong to all, and not to the few. The office lad and shop lass, the miner and the factory worker have equal right to traverse the heights and dales from one Youth Hostel to the next, and should be encouraged to do so, and take the ways of safety."

William Palmer never forgot his roots.

Late on in his life, he recalled an incident that happened when he was walking over Dunmail Raise which was to make him something of a hero to a band of Gypsies who were camped along the roadside.

Ragged children ran out from the camp to beg from passing coaches, one lost her footing, slipped and fell in the path of an advancing team of four horses. William Palmer, who was close by the situation, realised the potential danger and swept the child off the road to the safety of the grass verge. The coach passed on, to the accompaniment of muttered oaths from the coachman, leaving Palmer to be immediately surrounded by a group of excited Gypsies. At the centre of this jabbering crowd, little of which he understood, he was gently pushed towards a tent. There sat a small, bent and aged Gypsy woman, as black as the pipe she smoked. A younger woman was beckoned forward to interpret the old crone's words.

William was told he had performed an extraordinary feat in saving the life of a very special person, for she was the seventh child of a seventh child seven times over. Gold coins were produced from a red cloth bag and offered as a reward. William refused as he felt the act had been nothing out of the ordinary as far as he was concerned. He pushed the gold aside as graciously as he could, protesting that he had done nothing to earn it. He did agree however to allow the old Gypsy to tell his fortune.

"You may hope to live long for the days of ill health are behind you. You may hope to rise in the world, but it will be slowly, for you have too much pride and will not bend where you should. You never will be lucky in money matters, picking up money which you have not sown, but you will never be without money or dishonourably in debt. Whatever you want, you will have to work for, but you will get it in time.

Hills and rocks and mountains, mountains and rocks and torrent sides will be your pleasure and your fortune, but not for gold, not for gold."

As his life unravelled, it proved a fairly accurate prediction.

William Palmer's personality was shaped by his intimate involvement with the countryside and its people. The demands of city life and ambitious strivings necessitated the mantle of success. But his responses to the opportunities and challenges that life presented, provided him with his greatest joys.

How he fared is the rest of the story.

*Coaches and pedestrians
on Dunmail Raise.*

Chapter 3 - THE WALKER

William T Palmer started his fell walking career at the early age of five, when he was taken by his brothers to a tarn high up in the fells that overlook his home village of Bowstone. For some people childhood memories are difficult to recapture, but for others, there are moments of special significance that continue to remain clear in the mind. To William Palmer, the memory of that early childhood walk stayed with him, although he recalled that it was a weary trudge over what seemed to be endless miles of rough moorland. The long journey made him hot and tired; his bare legs were scratched and bleeding from the unsought for attention of sharp, dried heather stems. When he finally reached the tarn, which was the destination for an afternoon of youthful fishing, the reward of its sparkling water in the bright sunshine, washed away all discomfort of mind and body. It was not only the dark mountain trout that were hooked on that quiet sunlit afternoon. As a result of that first experience among his native hills, William was trapped by the delights that the fells can bestow on those who take the trouble to seek them out. At such an early age he made no conscious decision to take up fell walking, but the enjoyment of the day ensured that the seeds of enthusiasm were sown to come to fruition in later life.

They were nourished a few years later, when the "wise man" in the village allowed him to borrow a guide book to the Lake District. It contained maps and illustrations that showed there were wider horizons to explore than those in the immediate vicinity. That first guide book stimulated his idea to wander over, and explore, the whole of the Lake District.

His early explorations were devoted to the fells and valleys that were virtually on his own doorstep, but as he absorbed more information from the tightly printed text of that old guide book, it opened his eyes to the fact that there was even more excitement to be found further afield.

He grew up during the period and the economic climate, when money was scarce among the local population generally, and in the Palmer household in particular. The wages that his father earned from his work at the local paper mill were occasionally supplemented by helping out with shepherding at lambing or dipping times. Other casual farm work came his way when it was available and helped to fill the Palmer purse. The financial demands to feed and clothe a growing family left nothing to spare for the pursuit of fanciful interests of one of its younger members.

In the 1880's, there was no organised system of cheap, public transport to take the would be traveller around the region. It was possible to make journeys by train or horse drawn coach, but these were irregular and expensive. By comparison, foot travel to a young and healthy lad was immediately available and cheap. Thus, when William Palmer began to systematically explore the Lake District, it was by walking; for him there was no other way.

As a boy, he was forever wandering through woodlands and fields, and as he did so, he explored the quiet corners of the countryside near to his home. As he grew older, he roamed further and further away from his home village, venturing into, what was for him, unknown territory, as his horizons widened and his interests deepened.

Like many of the other village lads, he helped out with jobs on some of the local farms at nights and weekends, for there was always the need for a strong young lad to help out at busy times. In this way, he slipped easily into the traditional Palmer role of shepherding when he left school. His jobs took him into remote fell areas to help with gathering sheep that were scattered over miles of fell land. On other occasions, the animals had to be driven from one valley to another, as they were returned from wintering grounds to their summer grazing. The distances that he walked were often in excess of twenty miles, but this was nothing to a young shepherd lad. The pace was leisurely and he had the company of sheep and dogs and men. The time spent on the walks passed easily, as the distance was absorbed underfoot. He made a mental note of landmarks, he watched the activities of wildlife and he noted the changes to the countryside that the seasons brought.

His first long walk which he undertook for his own interest, rather than walking with the flocks, took him to Mardale. The remains of the former Lakeland hamlet of farms, church, cottages and inn is now submerged under the artificial lake of Haweswater, in the valley separated from his home village by the encircling ridge of the Kentmere fells. William reached Mardale by way of Long Sleddale and Gatesgarth Pass. His path took him over farm land, alongside sparkling streams, and across rough fell sides. When he reached Mardale, it was to discover, as had others before him, the existence of an enchanting valley with farm houses snuggled under fell sides. There were sheep on the fell and cattle grazing by lakeside meadows. It was an idyllic setting where bird's sang among honeysuckle hedgerows. There he found the twin focal points of the valley's life in the old church and the Dun Bull Inn where farm and hunting folk gathered. It was a valley with arcadian qualities and one that he left with regret. His view of the valley was vastly different sight from that seen today; for almost fifty years after William's walk, the valley was flooded to create the giant reservoir that quenches the thirst of far off industrial cities.

William returned home over Nan Bield Pass and down the Kentmere valley to complete the round trip of about thirty three miles. The memories of that first visit stayed with him throughout his life and the journey to the valley of Mardale remained a favourite walk, and was often repeated, although with the introduction of slight variations to the route. Sometimes he went by way of Shap, or the High Street range, or he even extended his journey north, to include a visit to the market town of Penrith.

These long country walks gave him the opportunity to develop his boyhood interests in birdlife, fishing, following the packs of fox or otter hounds, collecting nuts and blackberries from the hedgerows, and bleaberries from the fellsides. He thoroughly explored the valleys of Kentmere and Longsleddale, while the rounded grassy contours and hidden clefts of the Shap Fells and the Howgills yielded their secrets. He fell easy prey to the temptation that is proffered to the curious walker, to go just that little bit further

Mardale,
as William Palmer knew it.

Mardale as seen in the dry summer of 1975.

along a lane, to have a look over the next hill, or round the next corner. He was ever eager to scramble up to the next stone wall, or peer into a stone barn, before making up his mind to turn back and head for home. As he grew older, his enthusiasm for his former boyish pursuits, which had often been carried out in the company of other village lads, was overtaken by his passion for rambling for its own sake. From this, he gained great satisfaction from reaching the summit of a mountain, which to him was a tangible acknowledgement that he had succeeded in achieving a definite objective. William Palmer was a person who liked success and this was especially so if it was attained as a result of his own efforts.

The knowledge of natural history and country lore that he had accumulated for himself, stood him in good stead on his roamings. His walks were enriched and enlivened by his understanding of the ways of the countryside and its wildlife. Holes in banks and woodland floor denoted the presence of rabbit, vole or badger, while the emergence of a sharp, pungent scent indicated fox or marten. The fragile remnants of bone that he prised from the tightly packed fur of pellets scattered below an owl's roost, told the story of the night predator's hunting success, while slot marks at a muddy hollow revealed the secret of passing deer.

His enthusiasm for rambling set him a little apart from the rest of the lads in the village, but his far ranging travels aroused much interest and curiosity among their elders. His growing reputation as a long distance walker was greatly appreciated by the older village people. "Where hasta been today Willy" was often the eager greeting that he received on his return from a walk; for Willy was frequently able to bring news of friends or relatives who lived some distance away, and rarely met.

Like many others, before and since his lifetime, William Palmer discovered for himself that there is really no other way to get to know the Lake District, than to walk it. Motorists, or passengers who are restricted to the time table of a motor coach visit, only see the superficial face of an area whose beauty is more than skin deep. They travel along a narrow corridor of grey asphalt, forced to stay on the roads that skirt the edge of a world full of charm, mystery, romance, excitement and great beauty. For them, the Lake District is always viewed from a distance; it is never within reach. The way to get to know the Lake District is, in the local vernacular, "To git a hod of it."

Take a heather slope by the scruff of its neck and wobble on its uncertain stems. Cling to rough grey rock, inching fingers over gritty ledges while a foot is firmly jammed in the embrace of a rocky crack. Lie on a bilberry covered slope on a hot sunny day, where the distance of blue sky is paler than the purple stains of freshly picked tangy fruit. Venture, with trepidation, among the shattered heaps of quarry or mine waste and marvel at the men who, as long as five hundred years ago, burrowed like moles, in its underground darkness. Drink from an ice cold splashing beck; nibble the fresh lemon taste of wood sorrel, listen to the deep throated cronk of the raven; the bird, that above all others, is Lakeland's own. Watch the patterns of changing life that the season's bring to the hill farmer, the lambing, gathering, dipping, clipping. Visit a sports or valley meet, but keep silence; listen to the dialect , and you will enter their world.

Let your senses have full rein; taste it, feel it, smell it, see it, but above all, experience, respect and love it.

That was William Palmer's way, and his achievement in walking over virtually every part of Westmorland and Cumberland, combined with his specialist knowledge in other disciplines, made him one of the great authorities on the Lake District.

He started his explorations of the Lakeland hills, before the vast array of guide books marshalled themselves into pristine ranks along the bookseller's shelves. The great guide book writer, Alfred Wainwright had not even been born. When that event occurred, on the 17th January 1907, William Palmer had just recently become a member of the Fell and Rock Climbing Club at the age of twenty nine.

For his guidance on walking expeditions, William relied on the Baddeley guide books, with their tiny, close printed text. To this, he added the local knowledge that he gathered from the people he met on his walks. There was word of mouth from other travellers, that even included "gentlemen of the road", who, because of his disreputable appearance, mistook him for one of their own kind. They told him where he could obtain free, or friendly lodgings for a night, and although the information was mentally stored with thanks, it was rarely put to the test. His main ally in route finding was his linen backed, Bartholomew's 1 inch scale map. It's dotted lines, indented contours, and high ground shading, helped him to plot his routes.

Initially, as he moved further from his home district, he stayed within the security of the recognised road system. In the 1880's and 90's, these were not yet of the hard surfaced type that are accepted and occasionally grumbled about, in the Lake District today. In the latter part of the 19th century, the surfaces of these roads were stony and dusty in summer, and puddled with the muddy hollows and ruts of winter. When off on his travels, William's personal appearance altered according to the season, for when he walked these roads in summer, he was likely to be covered in white dust, or in winter, splattered with mud thrown up from the hooves and wheels of a passing horse drawn coach.

As he became familiar with the topography of the eastern fells, which border the lakes of Windermere, Ullswater, and Coniston, the further hills around Keswick, Buttermere and Ennerdale beckoned. Initially, he had no resources to avail himself of coach transport to reach them, the only way to get there was to walk further distances than those to which he had become accustomed. William Palmer was a great believer in training his body for long distance walking. He walked regularly every day, and gradually built up the distances that he could comfortably achieve. He aimed to be fit enough to walk fifty miles without a stop, and if necessary, he was prepared to take twenty four hours in which to do it. To the young Palmer, a day actually meant twenty four hours, it was not just the span of time that includes hours of daylight. It was his claim that he could "walk the clock twice round"; for him, time was too precious to be wasted.

In this way, he trained his body so that the walk to Grasmere became the normal way of getting to the start of a walk. For most people, the twelve miles there and twelve miles back at the end of the day would be a long enough walk in itself, without expending further energy on the high ground. For a while, William made Grasmere his starting point for walks in the fells around the village, which evoked the comment, "there was many a grim, sweaty, footsore mile" as he returned to his Bowstone home. His exploration of the fells accelerated, and once he had sated his appetite on Helvellyn, the Dodds, and the ridge from Grasmere by Silver Howe to Legburthwaite, he made Keswick his next objective. Just as he used Grasmere as a base for exploring the central fells, so Keswick became his base for the exploration of the northern and north western fells.

A typical walking day for William Palmer began when he left his home in the early hours of the morning for the march out to Keswick. From there, he explored Blencathra, or the Skiddaw forest, or even the Newlands fells before setting off for home when the light faded from the sky. Often he made his return journey either down Langstrath and over the Stake Pass to Langdale; or by way of Greenup Edge for Grasmere, from where he completed the familiar twelve miles journey back home to Kendal.

Although much of his walking was done alone, he was quite happy to have a companion with him on his long walks, but although he persuaded virtually every lad in the village to accompany him at some stage, after each occasion, the response was inevitably the same; once was enough.

"Usually I had to walk alone, for after one experience each lad of the village seemed to drop out."

Youthful hero worship of two notable 19th century walkers fanned the flames of William's enthusiasm for walking, for as most young lads of today have boyhood heroes in the form of football or other sportsmen, William's heroes were long distance walkers. He wanted to model himself on their achievements and to follow in the footsteps of two notable men, George Borrow and the Canon A.N. Cooper. They were his idols. As a young lad, he read over and over again accounts of their exploits. George Borrow was noted for some of the lengthy walks he made through the mountains of Wales, which though not all that far in distance from William Palmer's Bowstone home, seemed unattainable to an impoverished Westmorland lad. George Borrow's book, "Wild Wales" was a particular favourite, and the influence of Borrow's writing and walking formed a model for Palmer's own.

George Borrow's walking career was coming to an end, just as William Palmer's was beginning; but in those youthful days of hero worship, the descriptions of his long walks, set the standard to which the young Palmer aspired. For other inspiration, he looked to the long distance walking of the Yorkshire clergyman, Canon A.N Cooper of Filey, whose feats of travelling on foot to such far distant places as Venice and Verona captured William's imagination. The Canon's walks were to other destinations all over Europe, and in this respect, his achievements were even greater than those of George Borrows. WT Palmer wrote, "I was too shy to write concerning my hero worship and aspirations and in any case, the Walking Parson travelled alone on all the long journeys of which I have read." Thus was the inspiration provided for William to embark on his own long distance walks which were in excess of seventy miles.

William Palmer as a young man.

The first of these was undertaken in the summer of 1894 when he was then just aged 17. He left Kendal at midnight and walked to the top of Dunmail Raise, which he reached just as dawn was breaking. From there he carried on to Keswick, seeing the grey town through the haze spread by the smoke from early morning fires. There he paused, just long enough to have a bite to eat from the bread and cheese that he carried. Suitably refreshed, he took the path out of Keswick by way of Spooney Green Lane that leads to the now popular tourist route up Skiddaw. Once he reached the top of the 3000 foot mountain, he had no alternative but to return the way that he had come, for he knew of no other route off Skiddaw at that time.

By noon, he had crossed the Vale of Keswick and was on the steep climb of the rough surfaced road of Newlands Hause, passing dismounted passengers of the horse drawn coaches as they lightened the load on the steeper sections. Buttermere was the turning point in his walk. He left the huddle of inns, farms and a tiny church behind him as he walked up the equally rough unmade road of Honister Pass, before dropping down to Borrowdale to turn down the lane to Stonethwaite. A further three miles up the Langstrath valley took him to the foot of Stake Pass which he climbed over the fell for the descent into Langdale, and thence home to Bowstone. The distance that he covered in this single journey was 85 miles, and the time taken was slightly over 24 hours. At the time, he described himself as "a stripling of seventeen, not stoutly built, poor in dress and poor in pocket." He left home with 1/9 halfpenny [about 8 pence] in his pocket, and returned 3d [1 penny] the poorer.

When he was out walking, he preferred to travel with the sun at his back, so that he would not be dazzled by its glare, or lose sight of the detail of the landscape that hid in the depth of shadow. But once he had completed a particular walk with the sun behind him, he felt it was unfair to pass judgement on its quality until that same area had been walked in both directions. Like many of the earlier guide book writers, which included Thomas West and Jonathan Otley, who used a small hand mirror to study the scenery behind them, William Palmer appreciated the importance of the retrospective view. Any walker with an appreciation of landscape soon realises that it assumes differing degrees of grandeur and interest when seen from different viewpoints. After William had walked a route in both directions, any that he judged to be of no particular enjoyment, had little chance of being repeated.

William Palmer was not only interested in walking to the summit of a mountain to mark off its ascent as some form of achievement; of equally great importance for him, was the overall exploration of the countryside. One way in which he used to do this was to choose a particular contour line on his linen backed Bartholomew's map, and follow wherever it led him. When doing this, he ignored the marked footpaths, but made use of the sheep tracks which follow the contours of the fell like so many concentric rings. He negotiated any difficult obstacles that barred his way as well as he could. These included high rocks, waterfalls, chasms and gullies. If he could tackle them head on, he did so, but on occasions, he was forced to make detours before he could resume his walk on his chosen contour.

On one occasion he got into slight difficulties while walking in this way, through doing a good turn for some fellow shepherds. He came across a group of men who were intent on clearing out a raven's nest. This was common practise years ago, when ravens were regarded as a great threat to sheep at lambing time. The shepherds felt that by destroying the ravens' eggs, or killing the young birds, they would reduce the numbers of potential predators and thus lessen future damage to their flocks.

In nearly every Lakeland valley, a bounty was paid on any proven dead raven or eagle, and a stout rope was kept for the specific purpose of lowering a man to the rocky ledge where these large birds nested. William was accustomed to helping out with such a task; after all, in the course of his own shepherding work he had often been lowered on a rope to help to rescue cragfast sheep.

His encounter with some shepherds who were engaged upon this familiar task was no new experience for him; instead of sheep, there were ravens to deal with. William's offer of help was much appreciated by the shepherds who tied him on to the rope and lowered him to the relative security of the nesting ledge. As with most raven nest sites, it was sheltered by an overhanging rock and William needed to swing from the perpendicular to land on a ledge. From his landing point, he was unable to see into the huge structure of the nest. He untied himself from the rope and made his way along the rock ledge to the heap of intertwined sticks and heather stems that formed the years' old nest. In the centre of the mound there nestled a hollow, which was lined with matted sheep's wool. The nest was empty; the young birds had already fledged. There was nothing further that William could do, other than make his way back up the cliff to regain the open fell and resume his walk. This proved to be more difficult than he had anticipated, for the overhanging nature of the rock prevented him from reaching the safety of the dangling rope and its promise of an easy haul up on to the cliff top. The only way of escape that was left to him, was to retrace his steps along the ledge, then clamber through the large, smelly and abandoned structure of the nest, to gain an exit crack in the wall of rock beyond the nest.

This, he managed to climb, much to the surprise of the waiting shepherds, who may have greeted his appearance with wrinkled noses. William described the raven's nest site as being a most unpleasant place. "The debris of fowl and other small animals in all states of decomposition indeed makes the place a morass of filth." When he resumed his walk, he took the earliest opportunity of ridding himself of the stench by plunging, clothes and all, into the first suitable stream. The vigorous walking that followed was sufficient to dry his clothes as he continued on his way.

When William gave up his employment in farm work to take on a job in journalism that provided greater prospects of satisfying his literary interests, its accompanying improvement in his financial position enabled him to buy a camera. He sometimes took this with him on his walks. It was nothing like the modern 35mm "auto everything" model that is in common use today. The camera that he used was a bulky, mahogany, brass bound plate camera. On one occasion he was stalking deer on the Nab, a quiet area near to Ullswater. A forester, who knew both the area, and the ways of the deer, had helped William to take up a relatively close position to the animals, which afforded him sufficient opportunity to expose two plates. Some time later, however, when the plates were developed the results caused him considerable disappointment. His efforts were wasted, as the plates had been fogged by light getting into the camera. The cause of the trouble was apparently, "after seasons of hard use among mountain crags and sea cliffs, after several falls from gale swept positions, the camera had developed an unnoticed defect."

William Palmer was not just a fair weather walker, "either you must walk in the rain, or be content to waste many an afternoon," was his philosophy. Indeed, on some of his long walks, he appreciated the effects of a cooling shower, when "like a daisy, he turned his face to heaven." His personal attitude towards walking in rainy weather was one of unconcern. Wetness did not bother him, but he was well aware of the dangerous effects of becoming chilled in such conditions. For that reason, he advised others against resting or taking shelter in cold buildings such as a church, when one was soaked to the skin. It was his opinion that as long as the body was kept warm with exertion, one's layers of wet clothing could be ignored. After all, he had had plenty of experience of severe wettings without coming to any harm. Even after a sudden storm had "left pools in the slack of one's pants" and "it was inadvisable to sit down", the discomfort was of only a temporary nature. It was quickly forgotten after a hot bath and warm, dry clothes at the end of the day.

"No-one prefers walking in the rain, but it's a better sport than pining indoors."

He took his walking seriously and ensured that he was fit enough to cover the great distances that he set himself by training during the week. He would walk to work, five miles there and five miles back at the end of the day. He varied his route to make sure that he encountered different types of ground underfoot. Walking through soft farmland, scrubby woodland and the cobble strewn way beside a beck were all ploys that he used to break up the rhythm of his steps, to help to prepare for rough ground on the fells.

Work and family commitments meant that he had little time to spare in his early walking days, and so was determined to maximise what was available to him. Once he actually reached an area that he wanted to walk and explore, he was reluctant to lose what had been hard won by his long walk from home. For this reason, he preferred to "sleep rough", rather than face the long walk back home to his own bed. A lack of money for commercial accommodation accustomed him to spending fine nights on the open fell, or if the nights were so wild and wet that he needed shelter, then he looked for a barn. He made the most of every opportunity to cover as much ground as possible, and helped himself by travelling unencumbered by a pack that modern walkers seem to favour. He carried only the minimum amount of food that was essential to survive. A hunk of bread and cheese, and a few raisins were enough to accompany the readily available supply of drinking water from the mountain becks. A ground sheet or blanket, to give protection from the rain or cold, was all that he needed in the way of "creature comforts." On some nights, he didn't bother to sleep at all, for bright moonlight provided the ideal circumstances to experience the fells under different and magical conditions.

To avoid the need for sleep, William Palmer began many of his walks at midnight. This gave him sufficient time to climb a mountain and experience the enchantment of watching the rising dawn from a summit. He never failed to be stirred emotionally by the beauty that emerges when the darkness of a night sky pales to pearly grey, and then becomes diffused with pinky gold; or at times, when glowing angry, is streaked with fiery red. To sit quietly alone and absorb the distant views as they crept from dimness into the reality of light, gave him the best possible start to his walking day.

Towards the end of the 19th century, the challenge of long distance routes within the Lake District appealed to some of the walking fraternity. Records of sorts were created over a roughly established itinerary, which included the summits of the Lake District's highest mountains. In August of 1895, William Palmer joined with two friends to attempt to improve the established time of 19 hours 38 mins for climbing Bowfell, Scafell Pike, Skiddaw and Helvellyn. The starting and return point for their attempt on this record was from Elterwater Common. Even though it was summertime the weather was not in their favour, for the morning was wet when the three men set off. The party soon became hindered by very misty conditions as they gained height. With the onset of rain combined with cold, the ascent of Scafell Pike from Bowfell was only accomplished with some considerable discomfort. During the crossing of the rough, boulder covered ground, Palmer, who was normally so sure footed, slipped and injured his knee. He reasoned that the nature of his injury would slow the team down and prejudice their attempt on the record, so he dropped out of the total round. The remaining two were then able to carry on at a good pace. William left them as they returned to Esk Hause, from where they descended into Borrowdale. While they were able to make a fast and unhampered ascent of Skiddaw, William painfully descended the roughness of Rossett Ghyll. He walked and hobbled down Langdale and over Red Bank to reach Grasmere: there he waited to meet his friends as they made their way down from Helvellyn. In spite of his injury, he was fit enough to act as pace maker for his two colleagues on the final stretch of the journey. This helped his companions to reduce the record time for the ascent of the four tops by twenty and a quarter minutes. This record was not held for long, for subsequent and better organised attempts made further improvements on the time.

William continued to enjoy his walking throughout all the seasons of the year, and during the winter months, appreciated that snow and ice brought fresh experiences and greater potential for adventure. When he was in his twenties, he embarked on a five day Easter walk in Scotland in which he expected "a certain amount of exertion." This tested him to the full. The route that he planned took him through the central Highlands from Killin to Fort Augustus. It was a strenuous journey that he had planned, for it involved climbing mountains, exploring glens and included a long march through the Larig Ghru to Speyside, from where the lonely route of the Corryarrack Pass led to Fort Augustus.

When he left home, the Easter weather conditions in the north of England were less hostile than those of the Central Highlands of Scotland, and William Palmer was soon brought to this realisation when he encountered conditions of almost Alpine quality in his ascent of Ben Lawers. Mist and deep snow, and a howling, unceasing gale confronted him throughout his climb. The wind was so strong, there were a number of occasions when it threatened "to separate me from my grip on the ice axe." By contrast, his descent from the mountain was a rapid and exhilarating glissade down a snow slope of a thousand feet, from where he eventually reached his farmhouse accommodation in darkness.

The wintry conditions persisted into the following day when he walked over the top of Schiehallion which also had a good covering of snow at the higher level. Once again when it came to the descent he recorded, "I simply chose a line of deep soft drift and cantered at any desired speed." For the next stage of his journey he was impeded by torrential rain in Glen Tilt, where his encounters with swollen burns and broken bridges made the going "downright nasty", before he reached Inverey. He finished his journey yet again in total darkness.

His fourth day was to cross the Larig Ghru, which is treated with respect by walkers who attempt the crossing even in good weather conditions. William Palmer had been led to believe "it was no great feat in April", and therefore anticipated no difficulty on the walk. Even before he actually set foot on the pass, a local gamekeeper's warning of the dangers did nothing to discourage him from his chosen route. The difficulties were not long in becoming apparent. Surmounting deep snow drifts, skirting cornices, avoiding mini avalanches, and sweltering under the reflected rays of a scorching sun was the price he had to pay for the experience of seeing what he described as, "floating shadows of purple, delicate lines of pale blue, glorious mouldings of cream and grey. But king of all lights, king of all shadows, was the luminous blue which collected under the snow cornices, the great waves of which hung over the steep slopes and threatened the corries."

On this expedition, the inevitable delays caused by the difficult weather conditions prevented him from completing his planned journey, and he failed to reach Fort Augustus. The extra time that he spent wrestling with "all the wickedness a naturally bad path can muster after a winter in which torrents have been using its bed for a ploughing competition," the puzzling complexity of the Rothiemurchus forest, and the lack of a suitably timed train to take him to Kingussie, forced him to cancel the last stage of his walk. His holiday was over for he had to return to England to start work on the following day.

Although he did not recommend the harshness of this experience to the average walker, William Palmer advocated the benefits to health and well being that are to be gained from walking in general, and fell walking in particular. He remembered that Canon Cooper, his one time distance walking hero, lived to the age of 93. William Palmer did however, warn against the extra degree of difficulty that walking over rough ground can bring to those accustomed to nothing more than a country lane. "A fell walk calls for more than speed and strength; vigilance of eye and foot must combine to cope with the ever changing level. There must be a certain hold back of power, to change as a flash, a stride into a leap, and walk into a run."

In later years, the famous guide book writer Alfred Wainwright made a similar reiteration of those words, but in a more succinct manner. His advice was to "watch where you are putting your feet."

As an original member of the Fell and Rock Climbing Club, William Palmer sometimes joined in the organised monthly Club meets, but he was never really happy when walking with a large party. His account of being involved in a winter ascent of Helvellyn with a group whose "capacity as individuals varies so largely and a good proportion will be comparatively weak walkers," indicated his apprehension. He wasn't too enthusiastic about the inclusion of ladies in this particular party, for his comment of "Ladies too, unless inured to similar work, are better left behind," registered his feelings. In the wintry conditions that the party encountered, he recorded that one unfortunate lady fell down exhausted. "Stimulants were applied and she was led back to Grasmere." Further difficulties lay ahead for others in the party when "heated with their toilsome ascent they were in the worst possible condition to resist the attacks of cramp." In spite of more "freely applied stimulants", [no indication was given as to what the stimulants were], the distressed members of the party were unable to respond, and snow holes were dug to provide them with shelter from the wind until a rescue could be effected. This was certainly not the sort of fell walking that appealed to William Palmer. He was much happier when walking alone, or with a few companions of like ability.

When William Palmer left his Kendal home in 1918 to take up his new life in Liverpool, it proved more difficult for him to continue his fell walking and climbing in the familiar surroundings of the Lake District. But the enjoyment that he lost was merely of a temporary nature, for journeys in the Lakes, were exchanged for climbing and walking expeditions in the Welsh haunts of his boyhood hero, George Borrow. The Snowdonian mountains were within an easy cycle ride from Liverpool , and the memories stirred by Borrow's book, "Wild Wales" provided William with ideas of new ground to explore. He organised some of these into week long expeditions that enabled him to roam and discover a different countryside to that of the Lake District.

The experiences of becoming familiar with a new landscape, and developing relationships with people of a different language, culture and folklore, enabled him to continue his practise of recording all manner of items of interest. Historical events, natural features, places to visit, as well as any interesting tales that he could glean from the local people, were all faithfully recorded and reproduced in six informative books about Wales, most of which went into many editions.

He quickly discovered that his Liverpool base opened up many more tramping and rambling opportunities, for expeditions to the Pennines and the moors of Derbyshire were all possible, while easy access to the rail network from a mainline station eased his journeys to Scotland, and they continued to assuage his appetite for mountain walking.

When he embarked on these expeditions, he continued in the pattern that he had established when walking in the Lake District, true to the self sufficiency of his character, he carried only basic food supplies. These were supplemented by obtaining what he could from farms or villages through which he passed. On a single day of walking, he hardly carried any extraneous weight, but for the extra demands of a week long expedition, he carried a rucksack that, when packed with all he required, weighed about 16 pounds.

His usual walking attire was a tweed jacket with shirt and pullover, plus fours which became "pendulous when wet", but in warm weather these were discarded in favour of shorts although he continued to wear his customary long woollen stockings and nailed boots. The discomfort and irritation of insect bites, nettle stings and scratches were borne with great stoicism. He also found that after a heavy soaking, the wearing of shorts reduced the actuality of dripping water over the well scrubbed stone floor of the kitchen of a farm refuge. He preferred the temporary discomfort of stings and rashes, to the harshness of a Welsh woman's tongue. A mackintosh was often worn when starting a walk from home in wet weather. He preferred a long coat, "for the extra warmth and comfort it gave on the journey home." He appreciated the fact that a coat didn't whirl round his head in windy weather as a cape was apt to do.

On longer expeditions, he exchanged the greater comfort of a mackintosh for a ground sheet which served the dual purpose of affording a temporary shelter, while still being able to assume the scant protection afforded by a whirling cape.

The long journeys that he made over the rough mountainous country imposed great demands on his equipment, and even more so on his clothes. During one ten day tramp in Scotland that was carried out in a period of hot, dry weather, his boots disintegrated almost to the point of uselessness. The rough stony ground proved to be too demanding, the boots eventually succumbed to the point of surrender, and to William's great inconvenience, the uppers parted company from their soles. When this occurred he was still some miles away from any habitation or village There was not even a farm or a croft where a temporary repair could be made, so he had no alternative but to deviate from his route in order to "nurse" his boots on a smoother surface than that which had caused their near destruction. Only the all embracing support of bootlaces on the outside, enabled him to be able to keep his feet on the inside, and complete the walk in relative comfort. To add insult to injury, he concluded that at the end of his walk, "he had not written a satisfactory line."

On another Scottish expedition that was carried out during the winter months, it was not his boots that disintegrated before his return home, but his clothes. He had only taken one suit of clothes with him, which he wore every day. The constant rough passage that they were given of soakings and dryings, scuffings and tearings brought about their ultimate demise, and William's somewhat embarrassment. On the last evening of his holiday when his clothes had almost fallen apart, the lady of the house where he was staying, " spent considerable time and materials drawing enough of the soaked garments together for my journey south." He added, "At the first large place, I had to buy a suit of ready mades before I dared go home."

His physical appearance while tramping in wilderness areas sometimes created alarm among the local people. He was frequently mistaken for a Gypsy or a poacher; and at times was greeted with mistrust and apprehension. On one occasion when he approached a lonely farm, children ran to their mother with cries that there was a kidnapper approaching.

He made a point of calling at farms or cottages to check on routes, and gain directions or the latest information about local transport that was available. Route finding in the

Lake District presented no problem to him, for it was an area that he knew so well, but walking in the wilds of Wales, Scotland or the Pennines was another, and different matter. But the loneliness of these places that he encountered in his walking expeditions, found him tramping across vast expanses of featureless, bog infested countryside, where there were few crofts or cottages; and fewer people to ask advice about tracks and paths.

He was always aware of his responsibility to the countryside over which he was walking. If he was in any doubt about his legal right of way, he visited gamekeepers, or land owners to obtain permission for access, and ensure that he would cause no disturbance of stock. Usually he was affably received, and went on his way with yet another addition to his stock of "yarns" that would provide more grist to the editorial mill.

He depended on a one inch scale map and the appropriate Baddeley guide book to keep him on the right track, but, on occasion, when he wanted "to travel light", the book and even the map were left behind. This practice landed him in difficulties on the Isle of Mull, when he planned to climb Ben More. He found there was no one in the vicinity of his starting point who could give advice about the route up the mountain. William Palmer was not a man to be easily put off from his objective, and so he resorted to the mountaineer's ploy of following a stream to its highest point. The one that he chose to follow terminated in a grassy corrie, to which many deer had moved to escape the low level irritation of summer flies.

Initially, there seemed to be no identifiable way out of the corrie, and William spent much time wandering about the boulder scattered bowl before he was able to establish a route that took him to the top of Ben More. Much time had been wasted in seeking out the route. This evoked a comment, "I would have sacrificed a three half penny stamp for possession of that special map." The significance of the stamp is not explained, but there was much more wearisome scrambling on scree and rock before the final summit was claimed. His descent from the mountain was also a venture into the unknown, which brought a sharp reminder of potential danger to a lone walker in an unfamiliar area, whose whereabouts are unknown to any, other than himself. "A casual wanderer could be missing for a long time," he reflected.

William Palmer's walks of exploration were enjoyed as much at low levels as on the fell and mountain tops. A number of his low level walks followed the ways of the Roman legions, as he walked their roads. The journey from Corbridge to Scotch Corner was one of the routes that he took; he may have experienced an uncomfortable passage as he walked the entire distance, much of it on a hard road surface, in nailed boots.

William enjoyed having a purpose to his walking expeditions and one of these involved a weekend of walking along, and exploring the wildest sections of the Roman Wall that stretches across the north of England, for a distance of about 80 miles. The execution of some of these expeditions was done in, what most folk would regard as topsy turvy fashion, and his Roman Wall walk is a typical example. When William Palmer finished work in Liverpool one Friday afternoon, he travelled by train to arrive at Chollerford in the evening. There he enjoyed a late meal in an hotel, and then he set off on his walk during the hours of darkness. His reasoning for this was that it gave him plenty of time to explore all the interesting parts of the wall during the early light of the following morning. Later in the day, after he had been awake and active for almost thirty six hours, he decided to find somewhere to sleep. "In more rugged days I preferred to search out a quiet nook in the timber. Spiders, gnats and midges matter little to the healthily tired youngster. However, convention becomes stronger with the years." The accommodation that he sought was not quite conventional; for William Palmer sought Bed and Lunch from an obliging landlady, rather than the more usual Bed and Breakfast. After a late and satisfying lunch, he was sufficiently refreshed to carry on his journey along the course of the Wall, walking throughout the rest of the day and night. The following day, he completed the walk at Brampton before catching transport for his return to Liverpool, for work on the following day.

On the occasions that William Palmer returned to Kendal to visit relatives, he always took the opportunity of walking over the former familiar countryside of his boyhood. He was often accompanied by one of his many young nephews on these walks. One of them, William Walker is now in his eighties, but he remembers a teenage expedition in the company of William Palmer. "It would be early in the 1930's. We caught the market bus from Kendal to Long Sleddale. We walked over the tops to Mardale and stopped at Bampton where the hunt was meeting." From there, they walked to Shap and caught the bus back to Kendal. "Uncle Will would be about 55 then." William Walker recalled that he did not find Palmer an easy man "to converse with" . He explained, "He had a racy way of speaking, sometimes he would talk in the dialect which could be difficult to understand, but he could name every peak or bump in the land that you cared to ask about."

Kentmere, the valley in which
William Palmer began his walking career.

Chapter 4- THE CLIMBER

"Any young person of fair physique, clean life and with a love for the mountains may aspire to become a rock climber," were William Palmer's own words. He added, "Strength, courage, keenness, go to make up a rock climber, and it is only by their personal limitations in these essentials that enthusiasts are graded."

It was a natural progression for William Palmer to graduate from fell walking to rock climbing. In his exploration of all the Lakeland fells, scrambling had been an essential part of his exploits, for he was not averse to tackling any surmountable obstacle that was in his way. He negotiated many rocky areas in order to get on the ridges, where he enjoyed the feeling of airiness and surrounding space that their narrow pathways can give. He experienced the awesome exhilaration of walking over the crests of plunging crags and rocks, that tumble down to valleys hazed by distance. In such circumstances, he appreciated the insignificance of man, when set in the timelessness of a mountain landscape; an awareness of his frailty and vulnerability.

From his earliest ventures onto rock, William Palmer showed that he had courage. He was not afraid of the yawning depths below an exposed rock face, although he understood the dangers they presented. He assessed the problems that had to be solved for a successful ascent of a rock climb, and in so doing, he attempted to work out a safe solution. He had confidence in his own ability, which complemented the natural sure footedness of one attuned to mountainous terrain. The experience that he had gained, as a lad, through following the foxhounds, made him familiar with the rough ground of the rock strewn slopes of the valley heads. It is in this sort of difficult, high bouldered terrain, that the fox seeks the sanctuary of an earth, or a "bink." Even today, the followers of the hounds have to be prepared to get their hands on rock if they want to pursue the chase to its ultimate conclusion.

William Palmer's shepherding experiences had also taken him to the rough ground of crags and gullies in search of any sheep that had grazed their way into a dangerous situation. A surefooted cragsman proved his worth in such difficult places, in persuading reluctant sheep to rejoin the main flock. Occasionally he assisted with the difficult task of recovering a cragfast animal.

One practice to effect such a rescue, was to lower a man to the ledge where the sheep was trapped. The difficulty was then to approach, catch and truss the frightened animal without causing it to leap to its death. Wrestling with a frightened Herdwick on a narrow ledge cropped bare of grass, with a towering rock face above, and space yawning below his feet, was part of William Palmer's learning experience.

When he was in his late teens, his fell walking expeditions took him close to the great crags of central Lakeland, but although he was familiar with their general topography, he had only regarded the expanse of their rock faces from the viewpoint of a rambler. Dow Crag at Coniston, Ennerdale's Pillar Rock, and the buttresses of Scafell had not yet felt the scratching of his nailed boots. At that time, he did not include the uprising grey expanses of slab and tower as ground for exploration.

But in the 1890's, before William Palmer had showed any serious interest in climbing the crags for their own sake, others had already awakened to that challenge. For some years, leading up to the latter part of the nineteenth century, a few mountain men had already explored some of the crags in the Lake District, and opened up routes to the relatively new sport of rock climbing. Many of these early climbers were scholars who came to stay at a valley farmhouse or an inn for a period of study vacation. They spent most mornings reading, after which they took to the mountains for a few hours of relaxation before returning for further study. One of the most notable of this group, and a frequent visitor to Wasdale Head, was Walter Parry Haskett Smith. He achieved lasting fame in the annals of mountaineering as the first man to climb the pillar of rock jutting from the flank of Great Gable, which was known as Napes Needle.

If William Palmer had not yet familiarised himself with the great crags of Scafell and Pillar, another Will, had become a familiar to those early climbers. Will Ritson, was the owner and landlord of the extended farmhouse known at that time as the Huntsmans' Inn, where many of the climbers stayed. The inn later became known by its present familiar name, the Wasdale Head Inn.

Will Ritson was also master of an assorted pack of hounds of various shapes, breeds and sizes. Following them in search of foxes that went to ground in the stony heads of Wasdale, Mosedale and Ennerdale had made him more familiar with rocky ground than most local men. This familiarity ensured that he was in great demand as a guide to visitors to the valley, eager to try out the new sport. But while Will Ritson was prepared to accompany them on to the fells in his role as guide, he was not over enthusiastic about taking parties onto rock.

"Nobbut a fleein' thing could git up theer" he was frequently heard to remark.

In the 1890's , William T Palmer had not yet become acquainted or involved with the "Wasdale Head" group of climbers, who were mainly responsible for developing new routes on mountains such as Scafell and Pillar. It is almost certain that the young Palmer was aware of the rock climbing adventures of those men that stayed at Wasdale Head, even though at that time he had never met them. The son of a village labourer, great walker though he was, had little in common with the scholars and gentlemen of private means, who virtually took over Wasdale's inn and Row Head farm as bases for pursuing their rock climbing exploits.

Most of the climbers that came to Wasdale, were from outside the old counties of Cumberland and Westmorland. There were, however, a few native Cumbrian men who had, by their outstanding ability and skill on rock, made their mark on the sport. One of the most notable was John Wilson Robinson, a Lorton farmer, who accompanied many of the parties of visitors as a guide. There was also George Seatree of Penrith who teamed up with Robinson to set many new rock climbing routes. There were the Abraham brothers from Keswick, who were famed for their photographic records of Lake District landscapes and rock scenery, as much as for their rock climbing skills.

Wasdale Head and Great Gable.

In spite of Will Ritson's slightly derogatory comments about rock climbers not being satisfied with the fells alone, it was to him that John Robinson and George Seatree came for advice, on the best way to climb Pillar Rock. Ritson's knowledge of the Pillar was not so much through his own first hand experiences on the Rock, as by his association with the Reverend James Jackson. This man, who lived at Sandwith, near Whitehaven, became known as the "Patriarch of the Pillar", because of his affection and affinity for the Rock. Sadly, he made one attempt too many and was killed while trying to make a birthday ascent of the rock in 1878. He was eighty two at the time.

The Wasdale Head climbers usually organised themselves into small parties where the less experienced were accompanied by a leader who was already familiar with a chosen route. Pillar mountain, the Scafells, as well as Great Gable, became playgrounds of adventure for these men. It was not unusual for a party to spend a morning working out an ascent of Scafell Pike, while the latter part of the day was devoted to Pillar. The usual way of indicating that a climb had been successfully completed was to leave the names or cards, of those who had achieved their objective, in a bottle which was left on the summit, for that specific purpose. An alternative was to scratch their names on a small piece of rock.

So although rock climbing was firmly established in the region, it was not until the first decade of the twentieth century that William Palmer was to make the acquaintance of these, and other notable climbers, and join them on Lakeland rock.

William Palmer's first major rock climb happened almost by accident. After he had achieved fell walking ascents to all the major Lake District summits, he set himself the challenge of reaching them all again, but this time by more difficult routes. One day, he was exploring the western face of Great Gable, using what is now known as the Climbers' Traverse, or the Girdle route when he came across the slender pinnacle of rock that projects from the mountain side. This was Napes Needle, a spectacular outcrop of rock, which Haskett Smith had first seen in 1882, and climbed in 1886. The rock had been much photographed and superb enlarged prints in Keswick photographers' windows, proved almost as attractive as the rock itself. William immediately recognised the Needle.

Haskett Smith made the first ascent some time during the period 26th June to the 5th July, 1886, while he was staying at the inn at Wasdale Head. In his own recording of the event in the Wasdale Head Visitors' Book no indication is given of the exact date that the climb took place. The entry is a masterpiece of understatement when set in the context of the importance that his achievement assumed among the rock climbing fraternity.

"A fine climb of the arete character may be found on Gable Napes. This arete is the right hand bank of the right hand of the two great gullies which are seen from the hotel. It is marked by a peculiar pinnacle at the foot of it. The pinnacle may be recognised, [till the next gale of wind] by a handkerchief tied at the top."

These words, written by Haskett Smith himself are to be found in the "Wasdale Head Hotel Visitors Book", now in the care of the County Record Office, Kendal.

Haskett Smith's achievement ensured that Napes Needle became a popular objective with the large numbers of climbers that subsequently made the inns and farmhouses at Wasdale Head their headquarters. But in spite of the interest that was being shown in rock climbing on Great Gable, the second recorded climb of the Needle was not made until three years later, in 1889. It was becoming standard practise that only new routes, or variations of established climbs were being entered in climbing record books, so William Palmer's own ascent of the Needle was only documented in chapters of his own books.

When he climbed the Needle, it was also as a solo ascent, as Haskett Smith's had been. But Palmer did not, at that time or any other, have the great man's acknowledged rock climbing skill or technique. What William Palmer lacked in those qualities, he made up in others of determination and courage.

Like Haskett Smith, William Palmer had no rope to offer any assistance or protection, while on his climb, for in the early days of rock climbing, the use of the rope was somewhat frowned upon. To try and get more adhesion with the footholds on the rock, in his ascent of the Needle, William even discarded his nailed boots, which he decided were not giving much purchase on the rock. He left them at the base of the Needle and did the climb in stocking feet. This proved to give him a much better feel for the narrow ledges of rock that aided his upward passage. In one of his books published in 1934, "The Complete Hill Walker", he recorded his feeling of panic and terror during the climb, when his foot slipped, and a leg jammed in a crack in the rock. He was alone; nobody knew where he was. There was no one to help or encourage him. He was totally dependent on his own resources to extricate himself from a dangerous situation. After a few moments of extreme fear, he managed to loosen his leg, while still maintaining a hold with his hands. Once free, he overcame further obstacles, and successfully completed the climb. More to the point, he was also able to descend uneventfully to the greater security of the path, regain his footwear, and resume his boot shod way off the fell in safety.

Inexperienced as he was, he admitted that the climb was not done expertly, and he acknowledged that he may have been foolhardy to attempt it on his own. But the thrill of standing on the exposed summit of the pinnacle, with its tremendous view to Wasdale immediately below his feet, made him determined to "make a proper study of rocks and rock climbing."

He also wrote of this experience in his first book "Lake Country Rambles" which was published in 1902. He described the scene from the summit of his achievement. The description of this sight will be familiar to all who have actually climbed the Needle, and may stir the imagination of lesser mortals who have not.

"Across the narrow valley are the tremendous buttresses and scree beds of the Scafell range; down to the left is the dull gleam of Wastwater, with the greener shoulder of the Screes; behind you, pile upon pile, rises the huge composite cliff known as Great Napes; while right and left are inaccessible looking faces of bare rock. Glancing straight beneath, the climber is astounded; he is clinging to a mere pin point like summit over a tremendous gulf, at the bottom of which, seeming blue and distant, is Wasdale."

The experience of this climb did make a sufficiently salutary impression on him, so that many years later, in 1934, he pointed out the dangers for inexperienced climbers attempting solo ascents. With a reference to his own youthful assault on the Needle, he pointed out the advantage of sharing the experience with companions. "It is possible to get well and truly jammed in such places; with comrades, it is a matter for laughter."

He took advantage of the opportunity to study expert rock climbing techniques, at close range, shortly after his adventure on the Needle. William had been walking and scrambling on Scafell Pike and made the descent to Mickledore to watch some climbers tackling routes on the neighbouring Scafell. A roped party was climbing Moss Ghyll. William watched from a little distance, studying every move that was made, and noting every foot and hand hold they used. As the party moved steadily up the rock, William followed, but kept some little way behind. He did not wish to intrude on the party as they climbed the route, but as he watched, he noted, repeated and copied, as far as he was able, the exact moves that the climbers made.

The party that he followed were still on the summit of Scafell when he successfully completed the climb, and for a few minutes he joined them in conversation. They assumed that he was another climber who was staying at Wasdale Head, but whereas they had the relatively short journey to return to the comfort of the inn, William Palmer had the long tramp home to Kendal. His route off Scafell was by way of Broad Stand, a descent that walkers are still not advised to use, but William negotiated it safely enough.

In a number of his books, he makes frequent reference to Broad Stand as being a dangerous place for walkers, and also advised against its use. He even made the suggestion that a notice "Trespassers Will Be Prosecuted" would be the only way to deter those that were foolhardy enough to ignore such advice. No such notice has ever appeared however, and advice against its use by fell walkers is still ignored. Each year, mountain rescue teams are called upon to bring down walkers who still attempt the route, through lacking the necessary skills to complete the climb safely.

Encouraged by his successful, if somewhat adventurous, early experiences on rock, climbing became second nature to William T Palmer. He became acquainted with established climbers, and with their help, was able to tackle some of the major Lakeland crags. There was a group of Kendal men who were becoming proficient on rock, and were beginning to establish a reputation for themselves as being safe and competent climbers. They were already known to William Palmer and he willingly placed himself under their instruction. In the interests of safety. Such experienced men were only too willing to take a relative novice on their rope, and teach him the skills required to become a safe and competent rock climber.

He learned to tie the necessary knots quickly and easily, until it became second nature; he was made to practise moves over and over again, under the stern and watchful eye of a tutor, until he had mastered them perfectly. Only then was he allowed to progress to the next stage of rock climbing proficiency, and eventually develop to a degree of expertise where he was competent to lead climbs, and teach others as he had been taught.

He believed in the value of training for rock climbing, as he had done for long distance walking. Each day, he worked through his programme of exercises to develop his arm, and shoulder strength through a series of exercises that involved pull ups and dumbell work. These "workouts" helped to develop great upper body strength. As a result of this training, he found that when he was on the crags he was able to work his way along rock ledges, hanging only by his fingers, with little or no support from his feet.

William Palmer was of stocky build; he was short and burly in comparison to many of today's lithe limbed rock climbers. But for Palmer, physique was of secondary importance to nerve. "A slight defect of nerve or muscle is apt to come out at dangerous situations under the tremendous strains common to the sport, and a person whose confidence in himself is weak should never attempt really difficult climbs. If a man is deficient in nerve, it is impossible for him to become a cragsman."

A few years after his first ascent of Moss Ghyll, he discovered that the leader of the party that he had followed was none other than John Wilson Robinson, the famed Cumbrian walker and climber. At a later date William reminded Robinson of the occasion, which the older man also remembered. Robinson told him that he would have been most welcome to "tie on the rope", and join the party in the interests of safety. He added that solo climbing of that route, for one so inexperienced, was a foolhardy and dangerous thing to do.

Around the turn of the century, rock climbing and mountain walking in difficult terrain had gained a bad reputation in the eyes of many members of the public. This was due mainly to the occurrence of a number of accidents that resulted in death or serious injury to climbers. It seemed to matter little, to the general public, that many of these accidents happened in the Alps, where most fatalities did occur and not in Britain. To the uninitiated, mountains were identified as places where the foolhardy came to grief. Reports of mountain accidents were given prominent coverage in the press. The following report of an accident, was typical of what could happen, and the way in which it was reported.

This particular incident occurred on Snowdon in August 1882, shortly after Haskett Smith had seen the Needle for the first time. The victim was a climber who had previously stayed at Wasdale Head.

The Coroner for Caernarvonshire held an inquest at the Quarry Hospital, Llanberis upon the body of Mr T G Dismore.

Mr George Norton, a Liverpool solicitor, said that the deceased, who was an experienced Alpine climber, and had before ascended Snowdon, left Capel Curig with him on Sunday morning intending to cross Snowdon and descend at Llanberis. When reaching the copper mines, at the request of the deceased, divergence was made from the ordinary descent to Llanberis in the direction of Crib Goch. After descending a gully, they came to a steep point which appeared so dangerous that the witness urged the desirability of retracing their steps as the weather was rather misty. He turned back up the gorge and was under the impression that Mr Dismore was following his example, when suddenly he heard a noise, and turning round, saw his friend disappear over the edge of a precipice. He uttered no cry, but the body was heard to fall with a heavy thud on to some loose stones. The witness got round the ridge of the mountain, and after a delay of half an hour found Mr Dismore lying all of a heap, having pitched on his head after a fall of fully 300 feet.

The brain was protruding, and death, it was clear, must have been instantaneous. Dr. Hughes, the quarry surgeon deposed that there were extensive fractures on the skull, the bones being quite loose, and the brain exposed. Both legs were fractured, and there was no doubt that death was instantaneous. In returning a verdict of accidental death, the jury appended a rider, pointing out the necessity when persons deviated from the usual frequented paths, of engaging a guide."

This sort of graphic detailed reporting, did nothing for the reputation of the sport. Mountaineering clubs, which were then in their infancy, accepted that they had a responsibility to combat the bad publicity that was being generated. Among the members of many mountaineering clubs, the discussion was of safety; satisfactory procedures of progression were decided upon, rules of practise were formulated, and club members were urged to adopt these recommendations in the realisation that it was for their own well being, and the general good of the sport.

The democratic organisation that was practised by responsible climbing clubs provided the opportunity for them to exercise a degree of control over the activities of their own members within the acceptable and safety conscious guidelines of the sport. It also gave the opportunity for members to share their expertise, resources and information which ultimately led to the production of guide books that recommended routes of safe ascents and descents of crag and mountain.

This was seen as an attempt by the climbers themselves to be putting their own house in order as far as safety was concerned, and made the public aware that there was a responsible element involved in mountaineering. This helped to improve the image of the sport.

Most of the already established climbing clubs such as the Alpine, the Scottish Mountaineering Club and the Yorkshire Ramblers had been formed outside the boundaries of the Lake District, although their members were regular visitors to the Lakeland crags. Members of these established clubs also made frequent visits to the rock climbing areas of the Welsh and Scottish mountains, and further afield in the Alps, but prior to 1906, there was no formally recognised climbing club based in the Lake District.

For over 25 years, the great Cumbrian climber, John Wilson Robinson had advocated the formation of such a club in the Lake District. He advised other local climbers of the benefits that would ensue; but it was not until 1906, that the Fell and Rock Climbing Club of the English Lake District was formed. It had an inauspicious beginning when five men from the Furness district of Lancashire, who regularly visited the Coniston area for "fell rambling", decided to set up a club after three of them had spent a splendid day climbing. They wrote of "the Great Gully, Doe Crags on that glorious Sunday and felt that we just must get our fellow workers of the week to come and realise how much they were in blissful ignorance missing."

They appointed themselves as secretary, treasurer and committee of the fledgling organisation and invited the Keswick mountaineer photographer Ashley Abraham to be their first President. After some persuasion, he eventually agreed. It was only a short time before others from the Barrow and Kendal area were approached to join as "ordinary members." William T Palmer was one of the Kendal group of climbers who made up the original members of the club, along with such famous names in the climbing world as George Seatree, Walter Parry Haskett Smith, W.C Slingsby and many more.

There were about sixty original members of the club, and one of their aims was to make it possible for novices to have the benefit of learning safe climbing techniques from more experienced and established climbers. To put this aim into practise, they organised regular meets that were held in different parts of the Lake District. These meets served a dual role, for not only did they provide the training sessions from which many successful and long lasting climbing partnerships were established, but they also became good humoured, social functions.

From the very beginnings of the Club's life, it was deemed essential that safety measures in mountaineering had to be encouraged, if the sport and the Club were to flourish. Although it was inevitable that mountain accidents still occurred , for not every climber or walker was a member of a recognised club, the organised groups did prove their worth by helping to set improved standards of safety and ability. As an ever increasing number of first ascents of gullies and crags were made, or variations of established routes were developed, these were recorded in detail, so that other climbers could follow in safety. In time, these records became formalised into published climbing guide books, where the grading of rock climbs into different categories of difficulty was introduced. This set a structured form of progression to improve the climbing ability of both novice and the more experienced mountaineer. It also acted as an indication to climbers of the number and variety of climbs that were encompassed within their own range of ability and expertise.

In the early years of the Fell and Rock Club, no evidence of mountaineering skill or ability was required of any potential member, but even if it had been, William Palmer presented an impeccable case. He had already established himself as a long distance fell walker of some standing and he had proved his ability on rock. From the time he joined the club in 1906, until he died in 1954, his membership was continuous, although there were periods when he was relatively inactive as far as Club functions were concerned. His commitments to work and family prevented him from attending the monthly meets on a regular basis, but in many respects, William Palmer wasn't really a "Meets" man. In 1918 however, he became a Life Member of the Fell and Rock, and retained his interest in the Club activities and continued to contribute articles to the annual Journal on a fairly regular basis. As climbing became more popular and there was an increasing number of applications for membership of the Club, William Palmer offered to resign to make way for a younger member. His offer however was refused.

The memory of his own early exploits in solo climbing , when he was unprotected by either leader or rope, together with the recommended safety policy of the Fell and Rock Club, may have had some bearing on what he wrote in one of his books. "So many

Members of the Fell & Rock Club 1910.
William Palmer on front row, 3rd from right.

Walking dress in the early 1900's.

people are impatient to do rock climbs without the delay of training, or the trouble of a proper tutor." In his book "The Complete Hill Walker", he explains in great detail how, by following a logical course of progression, a walker can become a climber. As a member of the Fell and Rock Club, William Palmer benefited from close association with established climbers. The responsibility that leadership involved, and the appreciation of tackling routes well within his capability, changed his former youthful carefree attitude to climbing that he adopted in his first solo ascent of Napes Needle.

He became very aware through personal experience, which is regarded by many as the best of all teachers, of the dangers that rock climbing can bring to the unwary. On a number of occasions, William Palmer was involved in rescuing cragfast climbers from a difficult rock face, or leading off scramblers who had become trapped in a gully. In each case, he recorded that those climbers who needed this sort of help were not members of a climbing club, but individuals who had tackled something far beyond their capabilities.

He wrote of a group of scramblers who persisted in their upward climb of a gully in spite of the warnings that he gave them to turn back. The inevitable happened, and he recorded that "there they got demoralised, and stuck." It was only by his use of a top rope, that helped them to make a traverse across slippery rocks, that the scramblers were helped out of their dangerous position. On another occasion, the climbing party with which William Palmer was making an ascent experienced "the rope going tight." He wrote that while "the sensation from falling from rocks is horrible; that of being stopped by the rope is worse." The resulting nasty scrape to his ribs was a mark that William Palmer carried for some months.

With the acceptance of a rope for safety purposes it became a recognised part of a climbers equipment, and knowledge of its correct use an established part of climbing technique. Almost 100 years later, the principle of "belaying", which in its simplest form means tying a climber to a piece of solid rock or a sturdy tree, is still basically the same as in Palmer's day. The equipment may now be much improved and refined, but the manner of effecting a belay is similar. The rope, to which a climber is fastened, is looped through another length of rope, or a metal snaplink that is secured round a rock, tree, or other firm support. The latter may even be another climber, providing he is safely anchored to another belay. In the event of a fall, providing the climber is securely "tied on", his weight is taken, and hopefully held, before the fall has gone too far.

The effectiveness of a belay was put to the test with a party of five of which William Palmer was a member. Part of the route crossed a ledge which was not considered to be of any great difficulty. The leader and second man went ahead to secure a firm belay to bring across two ladies who were third and fourth on the rope. William Palmer watched them across from his position as last man. The ladies ignored the leader's instructions to cross the ledge one at a time, and both moved together. There was a moment of horror as first one slipped, and in so doing, dragged off the other. Palmer's warning yell and grip on his end of the belayed rope, together with the leader's efforts, arrested the ladies fall before it went too far. "Luckily it was an easy course, and the other belays held the people, who were scrabbling and knocking off bits of skin against the rough rock."

It needed some rough persuasion, in the form of severe tugs on the rope, to jerk the dangling ladies into a realisation that they had to make an effort to help themselves. In such a situation, even a hand or foothold on the rock wall against which the two were swinging, was sufficient to take some of their dead weight off the rope. Eventually with assistance from the three men, they were to regain the ledge from which they had slipped, and although they were shaken, both were able to carry on. " We completed the climb, but the lady never forgave me for the strenuous jerks on the line which roused her to effort. I think the sum total of her contusions were hidden - mine were very gory indeed."

Looking back to the rock climber's equipment of William Palmer's day shows that it was much simpler than today's array of high tech gear. Nuts, "friends", chalk, and pitons were unheard of in those far off days, although an old time climber would use a pebble or stone to act as a wedge in a crack of the rock. What would they have thought of the singlets and the multi-coloured tights that now splash moving patches of colour over the old grey rock. It would have been as strange to them, as Palmer's description of his own clothing and equipment is today. William Palmer recommended,-

"A rough suit of tweed which will stand rain and stain of moss and lichen, with puttees which will prevent getting a bootful of scree, any sort of old hat or cap will do." Any sort of old hat or cap gave little protection from the ever present danger of falling stone or other debris that whistled down a crag, as a ledge or gully that comprised part of the route, was being cleaned out. A climber's clothing, if less than fashionable, had to be equally serviceable, and was of a similar type to that worn by a fell walker; although often it was little more than old discarded clothes.

As far as footwear was concerned, climbers of Palmer's generation had little choice other than to use heavy leather boots, which had to be properly nailed. There were many debates about the most suitable pattern to be used, and climbers devised their own favoured combination of nails. The principle of their use was "Tricouni type nails [hard metal] should be used on soft rock, and soft clinkers on hard rock." Most of the early climbers had their boots made by hand, and a well worn old pair was often used by the bootmaker as the pattern for the new boot. At the turn of the century, there were eight bootmakers in Keswick, while in Kendal there were forty. A hand made pair of boots, of good quality calf, could be purchased for fifty shillings, which, with proper care, and resoling and nailing, could last for up to five seasons.

There was even a recommendation as to which was the best rope to use. "There is only one rope for the novice, Buckingham's Alpine Club rope." In advertisements, this rope was claimed to be used by the "Leading Mountaineers of the time." Its production was later taken over by the firm of Alfred Beale, but it was marketed under the same name with a warning to "Beware of fraudulent imitations." This particular rope was distinguished by its three red worsted threads that were woven into its intertwined hemp. To add greater authenticity, "Each length is tied with a Red Tape bearing my name."

The drag of those ropes in the pre-synthetic days , imposed a heavy, backward pulling weight on the leader, as he made his way up a climb. But this discomfort along with the rope's recalcitrant quality when wet or frozen, was forgiven for the extra safety that it afforded. All hemp ropes were difficult to handle in wet or snowy conditions, and the advent of modern nylon was one of the great advances in the development of climbing technology that William Palmer was not to know.

It could not be said that William Palmer ever became one of the true "Rock Tigers." This term applies to those who were ever eager to force new routes up the major crags. These Tigers revelled in the overwhelming sense of achievement of being the first to lay hand or foot on previously unclimbed rock. William Palmer was sensible enough to realise his limitations on rock, and maturity brought wisdom and safety into his climbing activities. In spite of his ability and all round skill as a mountaineer, the name of W T Palmer only makes an occasional appearance in the annals of Lake District rock climbing, and he never aspired to travel abroad for any Alpine ascents. He did however make a substantial contribution towards establishing a new major route on Pillar Rock.

This was one that had defeated a number of the top climbers of his time. The route was the "North West angle of Pillar Rock" on which he attempted a first ascent in 1905 with Fred Botterill, the brilliant Yorkshire climber. William Palmer had almost managed to work out the puzzle of the route, and succeeded in getting to within a few feet of the top of the climb when oncoming darkness forced them to climb down. They were unable to see the final way off. A short while later, Botterill and Palmer met up with Lehmann J Oppenheimer, climber, artist and sometime writer, who had reconnoitred the climb from the safety of a top rope. Oppenheimer had also had made an unsuccessful bid for a first ascent, but without getting to the height reached by the combined rope of Palmer and Botterill. In Oppenheimer's opinion, this pair had done all the hard exploratory work. They were persuaded to try again, but this time they had the added advantage of Oppenheimer's knowledge of the top part of the climb to help them. For this second attempt, Palmer and Botterill were joined by a Dr. J H Taylor, while Oppenheimer and some friends took up their positions above, and adjacent to the climb to supervise their efforts. Advice and a safety rope were both readily available if they were required.

In spite of the pooled knowledge of the two teams, the climb was not completed in one sequence. Exploratory diversions led to difficult situations, and both Palmer and Botterill had to resort to the use of the safety rope before the correct line was established. During the course of the attempt, the weather deteriorated, the rock became greasy and dangerous, and the climb was called off. A retreat was made to the Wasdale Head Inn where the discussions continued, but the route had been established. "And so the way throughout was discovered and all that remained was to do the climb properly" wrote Oppenheimer in his book "Heart of Lakeland."

It was not possible for the men to make a further attempt on that particular climb until the following year when on June 6th 1906, Fred and Arthur Botterill, Dr J H Taylor, and Lehmann J Oppenheimer made the first ascent of the North West angle of Pillar Rock. Work commitments prevented William Palmer being one of that party, but Oppenheimer

Snow cornice on Ben Nevis

acknowledged the contribution that he had made towards their success in the previous year.

William Palmer's interest in rock climbing was mainly as another facet in the make up of the complete mountaineer. He appreciated that rock climbing gave him another option to reach the summit of a mountain, rather than to regard the sport as an activity in its own right. The rock and ice climbing skills that he developed on the crags and in the gullies of the Lake District, gave him the expertise to follow more difficult routes to the summits of the higher Scottish and Welsh mountains, other than those usually frequented by tourists.

Foremost in William Palmer's mind was the fact that he was a writer, and it was as such that he had to earn his living. Writing up accounts of his ascents provided him with much needed copy for a demanding editor who wanted something out of the ordinary, and if possible, flavoured with excitement. Of a visit to Scotland, he wrote, "When on the hills incidents had to occur; with Ben Nevis in its usual bad humour there was no lack of them. On one occasion, I witnessed at close, but not perilous, quarters a thundering avalanche on the north face."

His summer Scottish ascents on rock, tended to stay within the easy, moderate, and difficult gradings, but he was also a competent winter climber. His stamina and tenacity gave him the ability to withstand many hours of debilitating conditions of cold and wind. His first experience of a winter climb on Ben Nevis arose from a casual invitation to join a party who were also staying at his farmhouse accommodation. William accepted, as he was always ready to try a new experience. His only winter equipment was a light ice axe, which he had brought to assist him on icy tracks at valley level. The day for the climb up one of Ben Nevis' north facing gullies was bright, crisp and cold. It seemed to present ideal conditions for winter climbing. Some many hours later, William Palmer declined to describe what happened during that climb on the mighty north face of the Ben, other than to say that " it gave him a good lesson in respect."

The gully up which they climbed led from the Mhuillin Glen, the upper reaches of which were closed by a huge cornice. For no reason of which William Palmer was aware, a sudden avalanche filled the lower part of the gully, and cut off their line of retreat. The only possibility of reaching the summit of the mountain was to hack a way through the cornice with an ice axe, which was a dangerous and difficult task. In the restrictions of the confined space of the gully, it was only possible for one man at a time to chip away at the ice, which proved to be a slow and tedious business. Several hours elapsed before the leaders succeeded in clearing a way for the rest of the climbers to emerge onto the summit plateau. Palmer waited till the other members of the party, and their equipment had been hauled through the gap in the cornice. He had an uncomfortable completion to his climb. "Perhaps the squad was impatient; they hauled with great vigour, did not give me a chance of working my way clean through the binding snow. At one time I felt as though my body would be cut in two."

It was some twenty hours after setting off from their Achintree farmhouse accommodation, that they eventually returned; the bright moonlight aided their descent from the mountain. In spite of the cold, the danger, and the length of time the party spent on the mountain, the experience did nothing to deter William Palmer from embarking on further winter expeditions that involved snow, ice and night time climbing.

The name of William Palmer is not well known to the present generation of the Fell and Rock Club members, even though in some cases, their periods of membership actually overlapped that of Palmer's. Apart from his exploratory work in establishing the North West Pillar route, the name of W T Palmer is not associated with any other recorded first ascents. There is no Palmer's Gully, or Palmer's Slab, or Palmer's Crack to carry his name forward among emerging generations of climbers. Other than his contribution of a number of articles to the Journal of the Club, and his eight year term as editor, his name is rarely mentioned in the publication. Yet in his comprehensive exploration of the fells, there must have been many occasions when he was the first to climb some rock outcrop or gully; the significance of which was insufficient at the time to merit its recording in a climbing guide. His own writings indicate that although he was an extremely competent climber, and well able to look after himself and others on rock, snow and ice, it was his first love of walking the fells, to which he was most faithful.

William Palmer was a regular contributor to the Journal, which is still the annual publication of the club, and during the eight years that he was Editor, he recognised the need for a wider range in the content of the articles. He attempted to broaden the range of topics that included climbing at home and abroad, to "reserve a space for articles on fell walking." In the 1910 edition of the Journal, he wrote, "The charms of wandering along the unfrequented paths of the fells are apt to be overlooked. The flora and fauna of the fells have been neglected of late years." The following year, true to his word, the Journal carried a lengthy article on "Fellwalking in Buttermere." William Palmer held the position of editor of the Journal from 1910 to 1918, the last year of which he shared the responsibility of office with his wife. This eight year period was not without controversy and he acknowledged that his "editorship was freely blamed for encouraging the barefoot fraternity." Some of the older members, clinging to the traditional methods of climbing, felt that an apprenticeship should be served in heavy nailed boots before attempts were made to master more delicate moves on harder climbs, that were performed in stocking or even bare feet.

This attitude was answered, and allowed space in the Journal, by newer members who claimed, with some justification, that they were expanding the horizons of mountaineering with their new techniques, while still adhering to the safety principles promoted by the Club.

In his last year of office, Palmer could also foresee that the ranks of the Club were likely to be swelled by a potential influx of new members from ex-service personnel of the first world war. He anticipated, correctly as it happened, that they would be from different social backgrounds to most of the established Club members. He was very concerned that they should receive the same welcome, and comradeship that was traditionally afforded to new members. William Palmer was a man with his feet firmly on the ground and was clear sighted enough to be aware of the different attitudes and aspirations of men returning to their former backgrounds. Adversity had presented many of them with the challenge of leadership, which a previous era had denied. He hoped that when such men submitted their applications for membership to the Club, the committee would adopt Palmer's own broad minded attitude. While not advocating that "a less severe scrutiny of nomination forms" should take place, he pointed out that "social status is a matter of less moment to me than to many."

He resigned from the position as editor in 1918, which coincided with his move from Kendal to Liverpool, and his last four contributions to the Journal were published between 1939 and 1942.

Throughout his lifetime's membership, he appreciated and enjoyed the companionship of fellow club members. While he attended many of the Fell and Rock Club meets in the early days of his membership, this diminished over the years, until with his move to London in 1946, they ceased altogether. When he was actively involved in climbing, it was the Coniston based activities that remained his particular favourites. He was not basically a gregarious man and preferred to walk and climb with just a few "like minded companions." On other occasions, he was more than willing to indulge in the more humorous side of outdoor meets. At Grasmere, when a number of the Club members had gathered to climb on Deer Bield Crag, William Palmer was responsible for starting a "gastronomic competition", based on the contents of members rucksacks. For a short while, climbing activities took second place to a feast of culinary delights which marked the meet.

"MENU;- lime juice drops, apples, jam sandwiches, chocolate, pears, beef sandwiches, cake, army biscuit [Palmer demolished one, no one else would tackle them], ham sandwiches, bananas. Thrice or more repeated and still the supplies did not run out. We all had tea at Grasmere."

Some of the Fell and Rock Club members of the latter part of the twentieth century, who had dimmed memories of William Palmer, failed to realise that he had once been a competent rock climber. There was an awareness among them that he had written books on the Lake District, but to many, W T Palmer was a name only found in the back numbers of "The Journal."

Sid Cross, a former landlord of the Fell and Rock Club haunt of the Dungeon Ghyll Hotel, Langdale, and himself a skilful rock climber with a number of first ascents to his name, met William Palmer on a few occasions. "We talked of mountains and mountain people, mainly those from Kendal who were founder members of the FRCC. I found him a most interesting person to listen to, but it was some years later that I learned of his long distance walking with an Ambleside man named Poole."

Another Club member who remembered him from fifty years ago was the late Mrs. Phil Wormwell. Phil was no mean climber in her younger days for among her personal achievements are recorded nine ascents of Scafell Central Buttress, along with two ascents of the Matterhorn. She recalled that her first meeting with William Palmer was below Dow Crag near Coniston, in 1939. William Palmer, who was living in Kendal at the time, issued a standing invitation to Phil and her husband to call for tea whenever they were in the area. From so many years ago, she found it difficult to recall exact details about their meetings which took place over the next five years. "The question of the cup of tea was not important, we enjoyed listening to WT chat about the fells and mountains."

The conversation was not all one sided. By the early years of the nineteen forties, William had left his rock climbing days behind him, but he was still interested to hear what other climbers were doing. Of especial interest to him, were Phil's accounts of her ascents of the Matterhorn. Phil remembered him as a very kind man, quiet and retiring, but very interested in what other climbers were doing.

One reason that is given for the lack of awareness and knowledge of William Palmer among current members of the Fell and Rock Climbing Club, is the fact that its membership is now so large. Present members of the club admit that it is virtually impossible to know of every other member. The current membership is spread throughout the world, and British members have found it more convenient for those in a particular geographic area to meet at the "climbing huts", in smaller groups. The overall intimacy of the Club's early days, with its small but enthusiastic membership, has declined as the development of the sport has raced ahead.

Chapter 5 - THE CAMPER

Camping isn't everyone's idea of a perfect way to spend a holiday, but to W T Palmer, it was the best way. To those who have experienced the questionable pleasure of life in the open air, with only the thin canvas of a tent for shelter in less than ideal weather conditions, the joys of camping, may not be fully appreciated. This is especially so, when there is the need to persevere with life under a dripping canvas shelter, for day after dismal day, with no means of escape to something more comfortable. The teeming rain of a Bank Holiday weekend, or the extended misery of a wet week in summer, does little to encourage enthusiasm for the outdoor way of life.

In such circumstances, the camper is presented with sodden, squelchy ground, that oozes water at the impression of every footprint on luxurious green grass. It leads to slippery, skidding steps, as the grass gives way to mud. Rain trickles down the inside of a collar; wet clothes, wet bedding, wet boots, wet everything. This is familiar territory to the camping fraternity.

To many who visit the beauty spots of the Lake District, the Scottish Highlands, or North Wales with high expectations of being at one with the spectacular mountain scenery that a camping holiday can give, disappointment comes along with the rain. The mountains hide their heads. They lurk under the burden of heavy, grey impenetrable cloud. The camper gains little comfort in knowing that the very nature of mountainous landscape brings the inevitability of rain.

Spirits sag and fall, while personal comfort disintegrates as the rain persists. Unceasing in its monotonous patter and damping rhythm, the rain drops beat the fabric of a tent. It dulls the brain of the most hopeful optimist, who waits for the break of blue sky among the grey turbulence of cloud, secure only in the knowledge that it can't go on for ever.

In fine weather, camping can be a delight; in the rain it can be misery.

William Palmer was full of optimism in his book "More Odd Corners in English Lakeland" when he recommended, "A camper's paradise is waiting for the holiday-maker in the English Lakeland, provided that he knows how to utilise the sites."

The sites about which he wrote were far removed from the fields full of regimented ranks of gaudy tents, that is the lot of today's camper. No longer is it possible for camping to be the once carefree activity which Palmer advocated, and that once delighted a relatively small number of visitors to the Lake District. Local bye laws and regulations have, of necessity, been introduced to effect a measure of control on the number of campers that now pour into the Lake District, and other areas of outstanding natural beauty. Camping has become big business, a part of the multi million leisure industry that increases with every year.

The increased amount of leisure time available to people, the ease of access to National Parks from a motor way network, and the availability of modern equipment, all contribute to the thousands of campers and backpackers that visit the areas of natural beauty throughout the year. No longer is there a closed season, although many of the large camp sites turn off their facilities for the winter months. Campers are creatures of all twelve months of the year. The sheer volume of their numbers demands regulation into carefully controlled and managed sites.

When William Palmer wrote his words, advocating the pleasures of camping in the Lake District, it was at a time when they were directed towards a relatively small, but growing band of fellow enthusiasts who sought out their own private sites that could be found tucked away in a remote grassy glade, beside an easily accessible stream or tarn, or among a shelter of tumbled rock. Only in some of the remaining wilderness country of the Scottish Highlands or on remote off shore islands, is it possible nowadays to recreate the freedom that William Palmer enjoyed in his camping days.

His advocacy of camping as an ideal way of spending a holiday, or getting to know the countryside, was born of years of his own personal experience gained through camping out in all weathers, and at all times of the year. The comfort of his portable home depended on the seemingly frail qualities of tautly stretched, yet fine material to keep the rain, sleet, snow, and wind at bay. Yet at other times of the year, he had the compensation of being immersed in the freshness of a clear spring morning , drowsing in the languor of summer heat, or being absorbed in a landscape rich in Autumn colour. His all year round experience gave him a balance of appreciation and judgement, unlike the sufferings of the summer novice, whose initial enthusiasm diminishes with each midge bite, or is flushed down the drains with the onslaught of a succession of wet days.

William Palmer's predilection for camping developed from his boyhood days, when, apart from an occasional visit to relatives, camping with the village lads was the only sort of holiday he ever knew. Over 100 years ago, packaged, or pre planned holidays that ordinary folk now take so much for granted were unknown. There was little money to spare in most families for anything other than the necessities of day to day living. While it was common practice for some of the wealthy families who lived in the industrial heartlands, or among the commercial centres in the south of England, to take a large house in some beauty spot for a period of months during the summer, those of lesser means had to be content with what was available at home. For some, a day trip to the seaside, or the lakes, was a luxury.

William made his own holiday pleasures in the company of other village lads. Local farmers tolerated their nights of youthful adventure, when they slept out in the comfort of a bed of hay in a stone Lakeland barn; or in some rough, home made shelter of branches that was erected in a farmer's field. During the hours of the short summer nights, they experienced the excitement of sharing the darkness with the wild creatures. The human-like snores of an impatient barn owl; the inquisitive snuffling as a hedgehog blundered its short sighted way through the undergrowth, or the eerie cries of a hunting fox, were enough to test the nerve of the bravest lad.

As William Palmer grew older and travelled further away from home on his weekend walking expeditions, he was not averse to sleeping out on the open fell, taking shelter that was afforded by some rocky outcrop that he shared, on occasions, with a patient sheep. "It is a joy merely to rest in the arms of the everlasting hills. The spirit is too alert, too exalted to accept sleep; the tired body may rest on soft dewy grass, but the soul goes a-climbing.On the grassy ridge there is always a play of misty terrestrial light; the world lies below like a sleeping giant." If the weather that he experienced on these walking weekends proved to be unsuitable for sleeping under the open sky, because of rain, sleet or snow, then he turned to the less romantic shelter of his boyhood days and slept in any available barn.

It was not until he was much older that he became an exponent of lightweight camping. The financial rewards of his journalistic work enabled him to have some money available to buy the essential equipment that he needed. He was a great believer in buying the best that he could afford. He reasoned that this policy would prove to be more economic in the long term, than buying a less efficient, but cheaper item on a more frequent basis. He used a tent that was small enough to pack into his rucksack, yet strong enough in design and material to withstand the worst of mountain weather. It had to be light enough to be easily carried in the mountains, for he appreciated the advantage of camping high, if his holiday was to be spent among rocks and ridges. "With a rucksack kit of tent, sleeping gear, and cooking utensils, there's delight in nights spent by quiet llyns and in rocky cwms," he wrote. When he travelled in mountainous country, he was quite happy to camp on the high ground, for this gave him the advantage of a good and early start to his days on the tops. That sort of terrain was hardly likely to provide him with a comfortable bed and he did not have the luxury of the modern day sleeping mat, to insulate him from the cold, and eradicate some of the wrinkles in an unyielding mattress of a fell side. But in keeping with his general philosophy of life, Palmer's approach was to make the best of what was readily available. He would try to find a really good ledge of soft, smooth grass, over which he made a meticulous search to pick out, wherever possible, the surface stones. "It is remarkable how soon you get to sleep on the ground which is somewhat humpy, and where you have to squirm parts of your anatomy so that the worst bits of stone are dodged."

For preference, if William Palmer was staying in an area for a few days, he liked to choose a site near to trees and bushes, and in close proximity to a farm where he was able to get day to day supplies. Most of the camp sites that he used on his lone walking or cycling expeditions, and about which he wrote, were of that type but he became increasingly aware, and accepted, that in the middle of the twentieth century, when there was an upsurge of interest in the outdoors and a growth in the number of camping enthusiasts, legislation was essential. He regretted that this resulted in a formalised type of camping on licensed sites, which afforded the camper with little communication with "country folk, apart from the satchel carrying site owner on his rounds."

William Palmer was what would be referred to today as a "wild camper." He liked to get away from the beaten track into the quiet areas in which he could walk, climb or fish without being disturbed, or even noticed. "The hiker who carries in his rucksack his home, furniture and provisions should certainly venture into the coves of the Lakeland peaks when the climbers amble down to dinner there is peace, except for the bleating of sheep, the croaking of ravens, and the songs of the mountain larks."

The Forgotten Man of Lakeland

Before he was married, in September 1901, William Palmer regularly left his home town to spend camping and climbing weekends in different parts of the Lake District, and often, these weekends were shared with two or three companions. Occasionally, William abandoned the climbing to his companions, while he found interest in some other activity. In his first book, Lake Country Rambles, which was published in 1902, he painted an idyllic picture of a camping weekend that he enjoyed near to the Lake District village of Coniston. His companions left him in the early morning to make their way over the tops for their objective of the day, Scafell .

William planned to have a lazy day in camp. He did all the essential campsite chores that included tidying up, he washed the dishes, and even "swilled off the grass." Later in the day , an old friend arrived with replacement supplies of milk and eggs, and stayed to chat for a while. After his friend had gone, it was time for a swim in the cold, fresh waters of Coniston before he retired to his hammock for an afternoon nap. The hours passed on to early evening, when refreshed by his sleep and a light meal, he went for what he described as a "short, lazy ramble", before joining a group of local men in a fishing and eel spearing expedition. "Locals have great faith in eelskin as a bandage for sprains, and to obtain a fresh supply, our expedition was carried out."

It was not until late evening that his companions returned; but that was hardly surprising, for their expedition to Scafell was without the benefit of motorised transport. They had walked from Coniston to Langdale, up Rossett Ghyll to Esk Hause, before swinging across to Scafell Pike and Mickledore for their day's climbing on the crags of Scafell. Their return journey was over much of the same ground. Their hard day was in sharp contrast to the lazy day of ease that William Palmer had enjoyed.

His companions' walkout to a distant climbing ground was common practice in William Palmer's time, and he thought nothing of walking from his Kendal home to Ennerdale. The distance that this involved inevitably meant that the return journey was frequently concluded in darkness.

Bad weather conditions did not deter William Palmer from his camping life-style. The experience gained over many years had taught him to be prepared for any kind of weather, and at any time of the year. He used a type of tent that could be easily erected by one person, and more importantly, could be taken down in a matter of minutes and resited in a more advantageous position, should the wind or weather suddenly change. He was skilled in being able to recognise the potential value of using any natural shelter that was available. The most obvious and the type of which he made frequent use, were sheep shelters and stone walls. They provided him with ideal protection to shield his low pitched tent from wind and rain. After all, he had watched the sheep on the fells for many years, and noted how they used the sturdy walls to give them shelter.

In his explorations of remote mountainous areas, where there was little chance of being sure of accommodation, his lightweight camping equipment enabled him to be totally self sufficient. William Palmer could be described as one of the first backpackers. "If I have a preference for one type of tent user, it is the cyclists or rucksackers who bring out a tent and equipment of minute bulk and weight, set up their field - home, cook meals and depart before the next dusk, leaving neither litter nor damaged sods behind them."

He carried all that he needed inside, or fastened on to his rucksack. William Palmer favoured the use of a Bergen type rucksack, with its abundance of pockets and facility for strapping items of equipment to the frame.

On all his expeditions, whether it was in the mountains, or exploring the lowlands, he aimed to travel as lightly as possible. To do this, he only carried sufficient food to last him for a couple of days, for he restocked with milk, bread and other basic provisions as he passed through villages or by farms. This essential contact with local people, as well as being the means of replenishing his vital supplies, also gave him the opportunity to chat to local folk about anything of interest that was to be found along his route.

The weight of the equipment and food that he carried for a camping trip of about a week was estimated to be about sixteen pounds. "My camp kit, which includes everything needed for hotel and kitchen, weighs about 16 pounds," he wrote, and the items that he included in his list makes interesting reading and contain one or two that are foreign to a modern day camper's list of essential equipment. [See list of camping equipment taken from "The Complete Hill Walker" published 1934 and included in Appendix A.]

William Palmer found that his style of camping was an excellent way to get to know the local people. Even the courteous act of calling on a farmer to ask permission to camp on his land , or allow him access to a route, ensured that he had at least one contact in a strange area. With William Palmer's natural curiosity, that one contact was often all that was necessary to unlock the door that led to other local characters with good tales to tell. The visits that he made to village shops and post offices, although of necessity to replenish his supplies, also served another purpose, for while there, he was able to eavesdrop on the local gossip. Such shops were, and where they still remain are, focal points of a village community and while the conversation may have been between local folk, and of local matters, many were happy to share their chatter with a passing stranger. "I like companionship and can never understand a district without some help from the natives," he commented.

William Palmer became an expert on camping, sufficiently so to produce a book called "The Art of Camping", which drew heavily on the accumulation of practical experience that he had gained over many years. He continued to buy the best equipment that he could afford, and although lightness of weight was an important factor, lightness had not to be confused with flimsiness, for his equipment had to be strong enough to serve his purpose. His kit was comprehensive, and included, tent, flysheet and ground sheet; woollen ground blanket and eiderdown; teapot/kettle, frying pan and two other pans, cutlery, primus stove and fuel, tea, butter, [in a protected glass dish] sugar, night wear and personal gear. Some of these items would raise smiles among today's plastic carrying backpackers, but William Palmer was able to carry this equipment quite comfortably in a framed rucksack for over a dozen miles of hard walking. After that, the weight began to make itself felt, especially if he added to it the extra 10 pounds of his typewriter, which he sometimes took with him on his expeditions.

In those pre - environmental awareness years, he was very much ahead of his time in that he was conscious of his responsibility to the countryside. When writing about camping in books or articles, he reminded his readers of the courtesy of asking permission from a farmer or landowner, before pitching a tent on what may have been private land. He advocated that no litter had to be left, or damage done to the surrounding area. He even discouraged the time honoured practice of lifting turf to make a fire place on the ground. Instead, he suggested that the cooking should be done on a small primus stove, the use of which was growing in popularity during his camping years. William Palmer took pride in the fact that when he left a camp site, there was no evidence left behind that indicated that he had been there.

When he married Annie Ion, he was fortunate in finding a partner who shared his love of the outdoors and was keenly interested in the wild creatures and especially the plants of the countryside. She was quite happy to accompany him on his camping forays, which could only be fitted into William's busy schedule as and when occasions allowed. The opportunities had to be seized when there was time available. This placed them very much at the vagaries of the weather, for during an early springtime camp in the valley of Buttermere, they were subjected to a weekend of violent storms of rain and snow. Even William Palmer's camping expertise was put to the most severe test in these conditions. After many attempts to rescue guy ropes and tent pegs, keep himself dry and warm and tether down straining poles, he reflected in his book, "Wanderings in Lakeland" that, "perhaps it would have been wiser to remain indoors for another fortnight, and so await better weather, but somehow such a policy has no savour for me."

Throughout his camping career, there were the odd occasions when William Palmer's camping equipment suffered minor damage as a result of encounters with animals who collided into the guy ropes of his tent during their night time wanderings. He could tolerate the head down perambulations of the hardy sheep, or the blunderings of the fell ponies that he loved, but he had no time at all for pigs and goats and dismissed their worth very summarily. "Pigs and goats, heaven defend us from such abominations! The first has the merit that it produces the breakfast rasher, but for the other there is no good word at all."

Camping expeditions continued to be a regular type of holiday enjoyed by the Palmer family, but with the extra comfort required for his wife, and later, two growing daughters, his holidays took on the further degree of sophistication that the company of womenfolk demanded. "I admit that years back I walked through the dales, and often on a summer night did not choose to take quarters at all, but slept behind a wall, or in a plantation. That, however is too rough going for modern times."

William Palmer was able to claim that during his lifetime, he had camped in every one of the counties of England. Added to that were a number of Scottish, Welsh and Irish counties where he used to pitch his tent as the base for his explorations.

He was methodical in the care and maintenance of his equipment. At the end of each expedition, everything was thoroughly dried, checked for damage, repaired where necessary, and then repacked. At home, it was always readily and easily available so that he could set off on an expedition almost at a moment's notice. While most of his camping trips were planned with careful precision, occasions did arise when a sudden departure was necessary, but in all his planning, he allowed for alternative arrangements to be accommodated within the overall scheme of things.

When the Palmer family left Kendal in 1918 to move to Liverpool, William was fortunate that his first employer was Lewis Chesterton who was, at that time, the editor of the Liverpool Daily Courier. This was a paper for which William Palmer worked on a regular, but freelance basis. Lewis Chesterton was regarded as something of an eccentric by the Liverpool city folk, for every weekend, without fail, he and his family left their urban life-style, to go camping in the countryside of Cheshire or North Wales.

In Lewis Chesterton, William Palmer found a fellow spirit, for the two men also shared an interest in cycling. The result was that the Palmer and Chesterton families became firm friends, and frequently joined forces on weekends that were organised by the newly formed branch of the Camping Club of Great Britain and Ireland. Lewis Chesterton was a founder member of the Merseyside Branch, and he made it his business to introduce William Palmer to the activities of the Club. Lewis Chesterton's son John, recalled those early camping expedition days.

"W T immediately became part of that circle of what we called the hard campers." He recalled that although the Chestertons went camping every weekend without fail, the Palmers were not quite as regular. "I think that was probably due to the fact that he was away on his own expeditions gathering material for books or articles." John Chesterton added "We were all regarded as eccentrics. We used to leave our homes in the early hours of the morning with the cycles loaded so high, you could scarcely see over the top of so many packages."

He recalled that they were frequently subjected to catcalls and shouts of abuse from knots of lads and men gathered together on city street corners, who watched them with curiosity as they pedalled their way out of town. The sight of a number of breeches, or shorts clad men and women evoked whistles from many of them. "It was quite shocking for girls to wear 'rationales' at that time; I was quite embarrassed at the reaction of some lads in Wales who were whistling at my sister," John commented. The sight of a number of cyclists pedalling away in the early hours of the morning, heavily festooned with poles strapped to cross bars, and panniers fixed fore and aft, was no doubt an amusing diversion to those with no escape from the drabness of city streets.

A favourite destination of the Camping Club was the Delamere Forest in the county of Cheshire. On these club outings the camps were very much a foretaste of things to come, for they were of an organised nature with upwards of twenty tents all neatly pitched in rows. The cycles were stacked tidily against a convenient tree, for no chaining of wheels as a protection against potential thieves, was necessary in those days. This organised level of camping was a complete contrast to the free camping enjoyed by William Palmer when he embarked on his solo ventures, but he was a sociable enough character, and enjoyed the relaxation of these weekends with his family and friends. "A certain number of people possess an inheritant taste for the simple life. There are others who like it for health reasons or as an antidote to an unnatural civilisation," he wrote in the Liverpool Daily Post in 1925.

The Forgotten Man of Lakeland

In the early nineteen twenties, there began an upsurge of interest in the outdoor movement, as people tried to forget the horrors of the 1914 -18 war years. William Palmer, although not actively involved in military service, had found these years a great strain on his mental and emotional resources, as many friends from village, fell and crag were lost forever. He revelled in the quiet serenity of the countryside and the relaxation that camping weekends gave, when he could turn his mind from the horrors of war to the peace and beauty that was around him. Others also found that their escape from the towns fulfilled a need to re-enrich their war ravaged lives yet, being part of the camping fraternity enabled them to hold on to the comradeship and companionship that the years of shared adversity had developed.

John Chesterton's words conveyed the mood of change. "After the end of the first world war, there was such relief that it was over, people were looking for some way to relax, and bring something of value back into their lives." This new enthusiasm brought a change of the former mocking attitude that had been directed towards those already involved with the outdoor movement. Some of those lads that once stood at the street corners in idle laughter, now took to the roads and open spaces of the countryside themselves.

As the interest in the camping movement spread among people throughout the country, families like the Palmers and the Chestertons ceased to be regarded as eccentrics. No longer was their departure from the city a source of amusement, instead, they began to find themselves among the leaders of a great outdoor movement, as many branches of the Camping Club of Great Britain and Ireland sprang up in many different parts of the country. Camping, as an outdoor pursuit, was here to stay.

Eventually, William Palmer became the chairman of the Merseyside Branch of the Camping Club, and although he was not able to attend all the Club's weekend meets, he made sure he was sufficiently active in his camping activities to justify the position he held. He enjoyed the humour and the camaraderie that these weekends brought into his life, as much as the opportunity they gave for him to shake the city dust from his feet. Among the members of the Camping Club were people from all walks of life. They brought a freshness of spirit and made a contribution of different interests and enthusiasms to their camping community. The campers included business and professional men from the City; workers from the dockyards, and families for whom camping weekends provided an excellent alternative to more sophisticated holidays that were far out of reach of slender purses. It was among such people that William Palmer was able to enlarge his circle of contacts. He loved to talk with people and listen to their tales, and these opportunities provided him with more journalistic material for his books and magazine articles.

Some of the tales arose out of odd mishaps that occurred, for as anyone who has ever camped is aware, there are always disasters that seem to lurk round the occasional corner. They seem to be major incidents of great despair when they happen, but with hindsight, can be turned to good humour, and laughed about for many years hence.

These were the tales for which William Palmer was ever on the lookout. He was not slow to capitalise on some of the misadventures that his camping companions experienced. He enlarged the tale of a botanist who gathered a collection of specimens in the forest, only to find they had been commandeered for a herb pudding. On another occasion, he described at length and with mischievous glee, the startled reactions of campers who snuggled down for the night only to find frogs and beetles hidden away in their sleeping bags. His well known aversion to some of the farm animals was given a good airing, as he wrote about a pig that rooted destruction among the guy ropes, or goats that ate the tablecloths. A foraging hedgehog was evicted from his tent which proved somewhat painful, "Bare hands and stockinged feet are scarcely the gear for battle with a hedgehog, but the brute was removed elsewhere with some show of haste and temper." He does not disclose whether the haste and temper was demonstrated by the hedgehog, or himself.

In spite of problems with an invasive hedgehog, William Palmer regarded camping as the best possible way of coming close to nature. The wildlife of field, river, wood and hedgerow had always interested him from the days of his boyhood. He quickly discovered that by life in a tent, especially if he was alone, provided him with the ideal opportunity to watch and study wild creatures at close quarters. It's amazing how readily birds and mammals will accept a small tent, even when suddenly erected, to emerge like a fungus bursting from the ground.

William Palmer was very fussy about the colour of his tent. It had to be a dull brown or green so that it merged into the background of the landscape. From the doorway of his canvas observation post, he watched the comings and goings of birds and small animals. Any nearby stream that he used for his water supply also came under close scrutiny; not for any pollutants but simply because he was interested in the amount of freshwater life it contained. He could identify most of the creatures that lived in that kind of habitat. A "pannikin" dipped into the stream usually emerged with more than just water. "The pannikin has lifted up wee frogs, tadpoles, newts even wee fishes. The water snails have been shaken off the reeds and grasses and sunk into the cup," he wrote in his book Wanderings in Lakeland.

Often he was just content to use the tent doorway to frame scenes of great beauty; slowly moving his eyes through the panorama and absorbing the wide variety of wild flowers and trees bursting into springtime bloom. "To those who dwell in tents by choice, May means light and warmth without glare and heat, a softened air and fleecy clouds in the blue sky. It means more and ever more flowers - and a beauty which almost takes away one's breath"

He wasn't always so fortunate in his choice of weather, but accepted that he had "to make his outdoor sojourns fit in with other conditions, and cannot pick and choose when and where I go."

At the organised camping weekends of the Merseyside Branch of the Camping Club, especially those held late in the season when the nights were drawing in, the Saturday night impromptu entertainment, by the campers themselves, of music, stories and recitations, helped to pass a pleasant evening. They were reminiscent of the Lakeland Merry Neets that William attended after a day hunting or shearing when the home spun quality of the entertainment was always a source of great pleasure to him. The two

Palmer daughters proved to be very popular with the rest of the campers at the Saturday night entertainments. They were both skilled musicians, especially Annie who composed her own songs as well as playing them on her violin, and she was always in great demand to lead the campfire singing, for a communal singsong round the campfire was the usual way of rounding off a Saturday evening.

The Camping Club's outdoor activities ceased for the winter months but the members continued to hold regular indoor gatherings to keep in touch with each other, and enjoy an evening of social entertainment. These activities were of sufficient general interest to the people of Liverpool, to ensure that reports of them were a regular feature in the Liverpool Daily Post. The fact that two officials of the Club, Lewis Chesterton and William Palmer, were on the staff of that particular paper, may have had more than a little to do with the coverage that was given.

Reading through the reports of these activities that took place over sixty years ago is a reminder of a gentler age. Their entertainment was home produced rather than commercial and relied heavily on audience participation. They were nights of eager chatter and friendly banter, when games were enjoyed with a ready, and harmless laughter. The intrusion of television into every day life, with its addictive powers, has not only broadened man's appreciation of the world, but in many ways it has destroyed family conversation and the fun of board games round a table. The type of entertainment that the Merseyside campers shared may sound quaint to the sophisticated ear of modern times, but it led to the enjoyment of simple pleasures, good humour and warm companionship.

These indoor gatherings that were held during the winter months gave the Palmer girls more opportunity to entertain the members with their music, Jean's cello was a somewhat unwieldy instrument to take to an outdoor camp and on those occasions, was left at home. But the indoor meets presented no such restrictions and the girls combined to play duets of popular music.

At the end of the evening's entertainment all the members still continued their practise of gathering round a camp fire, but for the indoor winter meets, the flickering of the leaping natural flames was replaced by the less romantic glow of a red electric bulb tucked away among a kindling of twigs and branches set in the middle of the room. William Palmer's own contribution to the camp fire entertainment was to revert to the dialect speech of his youth, to give recitations of Westmorland poetry. He was a life member of the Cumberland and Westmorland Dialect Society , and could readily turn to his native tongue. It added extra enjoyment to the evening although whether the poetry was understood when delivered in the broad vowelled, clipped speech, is debatable. "The success of the revel proved the Merseyside campers have the gift of chorus, song and that humour which is essential to the enjoyment of camps in the wilderness."

In later years, when the flush of his enthusiasm for cycle camping was restricted by the demands of increasing age and family size, William realised that a car was essential to carry the extra tentage and equipment that two adult daughters, a son in law and a grand daughter added to the list. But on these outings, however much luggage was packed into the car, William Palmer always ensured that there was enough room for his typewriter.

He never learned to drive a car, although at different times he did own two old Jowetts. The registration of the first was K4064 and the later one KF447. His unmarried daughter, Jean, acted act as his driver and among his books there are frequent references to "the lady who drives the car." He enlarged upon some of the difficulties they encountered in his book "Verge of Lakeland." He wrote, "My daughter who drives the small car often chides me for striking across ancient towns and cities in defiance of 'through routes' and recommended maps. 'You always get us into some twisted, steep, narrow, and pretty awful slum places which the Town Council whisper about when the Press is absent." Relieved of the task of driving, for which he showed no enthusiasm whatsoever, and having no personal problems in negotiating the twisted steep and narrow slums, William Palmer was free to concentrate on the work in hand, which was usually preparatory notes for yet another book.

A favourite camp site of the Palmer family was on a farm close to the Lancashire - Yorkshire border at Downham Bridge Mill. It was a place of happy memories for them all, whether they camped in a peaceful grassy glade with only members of the immediate family for company, or as part of a larger group with friends from the Camping Club. One of the big attractions of the Downham site for William Palmer was the friendly farm kitchen where many a tale was spun. Immersed in an atmosphere of home baked bread, the chatter of farm workers as they enjoyed a cup of tea, and the homely rattle of pots, reminded William of his boyhood days that were spent in many such a kitchen. In places like these, William Palmer felt at ease.

Although the family camps usually took place during the months from April to September, William was quite happy to take off by himself to camp out for a few days during the winter time. He recorded that he did so until just before the start of the second world war in 1939, when he was then in his early sixties. He spent a good many nights out on the fells, using a tent "which ventilated through the poles, and which could be anchored by stones on the outer sheet when the ground was frozen too hard to drive in metal pegs." The Peak District of Derbyshire provided him with many of his winter camp sites, but he pointed out that it took the real enthusiast to adapt to the dark and difficult conditions imposed by winter camping. "The campers must be hard folk and able to stand some strain."

In the 1930's, William Palmer took advantage of a more comfortable alternative to camping that was available to the outdoor adventurer, which was made available to them by the development of the Youth Hostels Association. This movement, which was started in Britain in 1930, had as one of its objectives, the provision of an opportunity to bring country holidays within the financial reach of young people who travelled by cycle or on foot.

The hostels which today range from purpose built accommodation to other converted buildings, were, and still are, planned to be about a day's walking distance from each other. Accommodation is simple, and hostellers can either eat meals that are prepared by the hostel staff, at a reasonable price, or they can cook for themselves in the

A harsh winter camp site.

Family camping 1934.

hostel's kitchen. Although modern technology has increased the comfort level in the hostels, the accommodation provided remains simple, cheap and clean, and each hosteller shares in the chores of keeping the place tidy.

Although William Palmer was never involved at a national level of the Youth Hostels Association, he was active in promoting the movement throughout the Lake District and made frequent references to its usefulness in many of his books. He supplied detailed information about their locations and suggested how they could be incorporated into the plans of a walking or cycling holiday. One of the most isolated of the youth hostels is still to be found at the head of Ennerdale in the form of the converted shepherd's hut. It huddles into the fellside at the foot of Scarf Gapand is known with great affection as the Black Sail Hut. William Palmer made reference to it in his book "Tramping in Lakeland" which was published in 1934. "The youth hostel is grandly placed, and I envy the lucky members who are able to use it. I am a pioneer member of the movement, and should exult in my opportunity, but I must confess that so far the place has not been visited under the new conditions. I am still in hopes that I shall go that way from Buttermere to Wasdale."

In the early days of the Youth Hostels Association, before the second world war, W T Palmer's enthusiasm for the movement in general, and the users in particular was not shared by many residents of popular holiday areas, where the provision of hostel accommodation was not readily available. William Palmer attributed this scarcity, in some districts, as being due to the aversion by local people for "the presence of youths and maidens who have practically no money." William Palmer had great sympathy with these penniless youngsters; he understood their problems only too well, for he had often been in the same position himself. He admired them for their determination to venture out under their own efforts; to respond to the challenge of exploring the countryside on foot, and the way in which they were prepared to depend on their own resourcefulness, in much the same was as he had done.

To many of the hoteliers in parts of the Lake District, Scotland and Wales, the appearance of these travel stained, purse strapped hostellers was upsetting. Owners and managers of hotels and guest houses were apprehensive that they would have a disturbing effect on their guests. They feared that in some of the popular regions, the appearance of such characters would "cause a shock to their ordinary guests and a deterioration in the class of visitor."

How wrong they were, for as the hoteliers were to find that there were customers enough for every type of establishment and the Youth Hostel movement expanded, diversified and responded to the changing needs of outdoor people. William Palmer advocated the use of hostels by young people wherever possible, for in just the same way as he looked out for the "Winged Wheel" logo of the Cyclists Touring Club that assured him of its recognised standard of value for money accommodation, so the familiarity of the triangulated letters of the YHA were eagerly sought by young people.

The years of the 1939-45 war, forced the temporary closure of many hostels when there were neither the facilities available, or the number of people with time to use them. Some hostels were able to keep open and they provided an escape and a well earned rest for many service personnel on leave from the war. They also provided holiday facilities for those who were still able to take a short break but of necessity it was under severely restricted conditions. It is interesting to look back to the conditions under which the Youth Hostels operated during the war years and William Palmer has done a service to the history of the movement by putting these restrictions on record, in his book "Odd Corners of the Yorkshire Dales." It makes fascinating reading for the hosteller of today.

"It is advisable to take ration books, and to carry extra butter and sugar in case there is difficulty obtaining these. Members should keep themselves informed of rationing regulations, and any change that may be made. Members wishing to do their own cooking may obtain, from any Food Office, emergency coupons which, when used in conjunction with the ration book, are available anywhere. Members are advised to be inside the hostel by 9pm every day. Identification cards must be carried as well as YHA membership cards. Talk of aerodromes, military, naval or other establishments must be avoided in all youth hostels at all times. Owing to difficulties of Wartime replacements, in practically every hostel, members are required to bring their own knives, forks and spoons."

The movement has changed dramatically during the 60 years of its lifetime. The comfort of modern day hostels equipped with central heating, hot water, bright, well equipped kitchens where chores are reduced to a minimum, contrast with the stark limitations of the war years.

William Palmer continued to enjoy his camping and youth hostelling until the latter years of his life. His camping needs were simple; "I still sleep on a down quilt, with thin ground blanket, and waterproof ground sheet between me and the earth." But with deference to his advancing age, he gave up his bed under the open sky, and made more use of the hostels whenever he could, not only for the greater comfort that they allowed him on his travels, but because of the companionship that they afforded in the common rooms.

He found them "convenient, with good company in the common room [though my grey hairs caused some raising of the eyebrows, followed by twinkling of eyes]."

Ron Winter a member of the Fell and Rock Club and a former member of the Cyclists Touring Club, recalled that while staying at the Keswick Youth Hostel, he occupied a lower bunk in the men's dormitory, while William Palmer used the top. Although the actual date is not clear in his memory, he estimates that it must have been in the early 1950's. At that time William Palmer was then in his seventies. Even from so many years ago, Ron's memory of William Palmer is of a great raconteur, and very good company.

Chapter 6 - *THE CYCLIST*

The words of the colloquial phrase of the 1990's, "On your bike" would have meant no offence to the teenage William Palmer a hundred years ago. In the 1890's, before motorised transport was readily accessible to the general public, William Palmer had almost as much enthusiasm for cycling as he had for walking and climbing.

There was hardly any other alternative means of transport available for the lad that wanted to travel, not only in his native county, but further afield to other parts of the north of England. The wild and lonely areas of the North Pennines and the Yorkshire Dales that he wanted to explore, were not so far away from his Bowstone home in terms of mileage, but when the only way of getting there was on foot, they could have been at the other end of the country.

William Palmer had become, sometimes painfully and often wearily, aware of the fact that although walking was cheap, it was intensely time consuming. To one who had only limited leisure time at his disposal, he had to make the maximum use of whatever was available to him.

Towards the end of the last century, the railway network was spreading its iron tentacles across the country, yet the recognised form of travel in remote rural areas was still by horse drawn coach. Kendal was a major centre of the coaching and carrying network, whereby passengers and goods were distributed throughout the north of England. Each day a regular service left the town, to carry passengers across the Lake District to the industrial area of the West Cumbrian coast, while others journeyed inland to the dales of Yorkshire and Durham. There was a major route through the Shap Fells for those coaches that made the journey to Carlisle and thence to Scotland.

The big drawback with coach travel, as far as William Palmer was concerned, was that their use, for even a short stage of a journey, was very expensive. The price that was quoted by some companies was as much as 1/6 for a mile. This put coach travel way out of reach of a young lad's pocket. Although the coaches adhered to a regular time table, especially when the mail was carried, they were infrequent. This, together with their strict time table, placed an irksome restriction on a would be wanderer even if he had had the money to pay for his fare. As there was no feasible alternative, other than time consuming walking, William turned his attention to the bicycle.

The first cycle runs that he made were on a heavy, cumbersome machine that he hired from a local blacksmith. He recorded his earliest cycling experiences in the following way, "I tried to learn on an old boneshaker built by some wheelwright; trips usually finished in the hedge or the ditch. The first pedal cycle on our road was a marvel and a show. Later I took the cycling fraternity rightly as the salt of the earth."

The garages that were opening, to tend the needs of the first motor cars, realised the potential business that cycle hire could generate, and they, along with some of the coaching inns and wheelwrights shops, provided a ready supply of machines for hire.

William Palmer eventually acquired his own cycle, which he described as being as equally heavy and cumbersome as those that he hired. The first machine that he owned was an old roadster; it was solid and substantial and it required a great commitment of energy and effort on the part of the rider to obtain the necessary propulsion to turn the pedals. He sat uncomfortably upright as his legs laboriously turning the heavy rubbered pedals. The riding position was such, that the maximum amount of his body area was presented to the wind. With such a 'sit up and beg", machine as they came to be called, there was no possibility of adopting a streamlined posture to reduce wind resistance. The resulting buffeting was inevitable.

At the turn of the twentieth century, bicycles were accepted for what they were, "solid tyred safeties", was the description given to them by William Palmer, but far from regarding them as cumbersome and unwieldy monsters, as a modern day cyclist would, he appreciated that ownership of such a bicycle had a great advantage; it solved his problem of transport. Owning such a machine enabled him to get further afield to continue his explorations of the countryside.

The independence it brought to him, enabled him to set off on his travels whenever, and wherever he liked. It was the means by which he was able to satisfy his enquiring and adventurous nature. He was quite willing to trade the long miles of hard pedalling, for the greater distance he achieved, and the greater freedom that it gave him. He was more than satisfied by the fact that cycling to reach a starting point of a walk required far less physical effort than making that same journey on foot. Cycling enabled him to increase his speed of travel to a destination, from four or five miles per hour when walking, to between fifteen and twenty miles per hour when cycling; he also had the added advantage of increasing his speed substantially when going downhill.

Of his early cycling ventures he wrote; "On our heavy machines, not ridden expertly and apt to break down, the pace was slow and laboured, the hills were rather a bore." His rides took him far away from Kendal to make Preston, in Lancashire a regular destination, and sometimes he extended the ride to Liverpool, from where the return journey home was made by train.

These journeys opened his eyes to the great variety of landscape that is to be found across the north of England. For a boy who grew up among the clean fresh fields and fells of the southern Lakes area, the grime and smoke of industrial Lancashire was something of a culture shock. As a boy, his early experiences of Lancashire had been gained when he stayed with relatives who lived in those parts of the county that once merged into rural Westmorland, and this had presented no great change in the rural environment. Cycling into the industrialised heartlands of Lancashire, gave him his first sighting of tall mill chimneys, narrow streets with brick built houses and smoke laden air. As he pedalled through the quiet Sunday towns, there was a strange silence, as freed of their daily routine of clattering clogs, the streets slept.

The flatness of the Lancastrian industrial countryside, made for easier riding than the hills of Westmorland, and this allied to the relatively good condition of the major roads, helped to compensate for the heaviness of his machine.

Some early cycles were built in a wheelwright's shop.

*Pneumatic tyres and lighter frames made life
more comfortable for the cyclist.*

Another of his favourite early morning rides was to cycle, from his home in Kendal, to Keighley in Yorkshire. On the way he passed through the market town of Settle. His usual programme was to leave home at five o clock in the morning arriving in Settle in time for breakfast and a shave in the barber's shop. Imagine modern day cyclists thinking of such a thing! Breakfast maybe, but a visit to the barber's shop?

After his break in the market town, William Palmer was then suitably refreshed, both internally and externally, to continue on his journey. The rough road surfaces of the Ribble valley with their sharp bends and steep descents, together with heavy horse drawn traffic, presented him with some problems. He often persuaded himself that it was safer to dismount and push. "I must insist that at no time would you have given me credit for neat pedalling; I fear that I only looked upon the cycle as a vehicle to get to this distant point and that beyond the range of walking."

One of the major problems he encountered was that presented by the movement of livestock. Regular sheep and cattle markets were held at convenient central locations, and these were a regular feature of many of the rural areas. In order to reach these from outlying farms the animals were driven along the same roads that were used by other traffic. Before the days of transporter wagons, the large numbers of slow moving animals caused unavoidable congestion for other road users, and presented a hazard to the cyclist. As William Palmer quickly found to his cost, any attempt on his part to weave a way among the moving flock or herd could easily result in retaliation on the part of a drover. All that was needed was a well placed stick pushed between the spokes of a wheel. This achieved the drover's objective of separating the cyclist from his machine, and caused the unfortunate rider to have yet another tumble on to the dirt splattered road. This was most likely to occur "especially if you were slim enough to be trifled with."

William Palmer's bike bore little resemblance to the sleek machines of today, although the component parts were not too dissimilar. His bike had a sturdy triangular shaped frame with upturned handlebars and a rubber bulb type horn was attached to the crossbar in place of the modern day bicycle bell. The wheels were fitted with heavy walled tyres and they slotted into substantial front and rear forks. William's first machine had solid tyres and although they were impregnable to the sharp stones, rock edges and thorns that found their way onto the road they provided little in the way of comfort. It was some years later that he graduated to the more comfortable, but liable to puncture, pneumatic tyres.

His early machine did not have the advantage of a multi gear system; he rode with what is known as a fixed wheel. The principle of this being that as a cyclist pedals, a large toothed sprocket, over which the chain passes, is turned; this drives the rear wheel, and thus propels the cycle forwards. Conversely, as long as the rear wheel rotates, the pedals continue to turn. Sometimes William Palmer found that his pedalling was involuntary, especially if he was on a steep descent, for then the pedals pumped his legs furiously up and down as they continued to turn. He found that one advantage of the fixed wheel system was that he could help the not too efficient braking mechanism of his machine, by attempting to hold the pedals back, and thus, attempted to slow down the rear wheel. 't was a dangerous ploy on a steep descent however, for he often found difficulty in actually keeping his feet on the pedals, let alone hold them back. On these occasions he preferred to enjoy the exhilaration of a downhill run as he swept down a steep slope with his feet on the handlebars in an attempt to escape the whirling pedals. Occasionally, one of the consequence of enjoying such heady, uncontrollable speed was that the rider was pitched headfirst into the nearest hedge. "There was a wonderful run past the well [Ebb and Flow Well in Ribblesdale], but on one of my last cycle rides I had to brake violently and come off."

Bumps and bruises, cuts and grazes were accepted as part of the hazards of cycling, but in those pre tarmac days, William Palmer was sensible enough to let wisdom prevail. If a very steep road with a rough and unstable surface had to be negotiated, then he found it more comfortable, and safer to walk.

When he progressed to the ownership of a lighter weight machine of a much improved design, he began to enjoy cycling for its own sake, rather than as an end to a means and shared his enthusiasm with others who lived in the Burneside area. A nephew of William Palmer, who still lives in the village of Burneside, recalled that while there wasn't a formally established cycling club in the village, "Uncle Will used to gather the village lads together and organise bike rides."

In the later years of his cycling career, when cycle design had improved still further, and more efficient lighting systems were available, he enjoyed winter and night time riding. In his early cycling days at the turn of the century, he found this sort of riding to be a somewhat hazardous affair. He quickly discovered that riding on ice was dangerous, and related that "rubber and ice make poor companions", with the result being that "rider and machine frequently part company."

Riding in the dark in those early days presented its own problems, for before electrically operated lights by a dynamo system were available, the only illumination of the darkened road ahead was from an acetylene lamp. Compare this account of maintaining an early 20th century cycle lamp to the ease of switching on a modern day electric dynamo system.

"An acetylene lamp will give trouble if you do not know how to treat it. If you are using it regularly, the chamber will have to be taken out once a week and cleaned with hot water and a stiff brush. The valve controlling the supply of water to the carbide should likewise be washed with hot water and lightly oiled. Never allow the flame to smoke; this forms soot which is liable to choke the burner."

This was the type of lamp that William Palmer used on his cycle, and if it was carefully maintained, he could be assured of an excellent light. When he did eventually acquire an electric dynamo system for his bicycle, the reassurance of a beam that illuminated the road on night time rides, gave him the impression of being cocooned in darkness, with only a soft yellow light opening the way ahead.

Nearly all of the cycling that William Palmer did was of a touring nature; he was never tempted to race or enter competitive events as he had attempted in the wrestling ring. Long days spent in the saddle inevitably brought some personal discomfort, but he did what he could to alleviate the problem. One piece of advice that he gave to fellow cyclists in an attempt to overcome this, was to use a broad saddle whenever possible. In those pre-synthetic days, a bicycle saddle was made of leather and regular care was needed in order to keep it in as good and supple a condition as possible to provide for maximum comfort. It had to be first of all washed with saddle soap and then given a good dressing with neats-foot oil. It was recommended that this treatment should take place at least once a year. The fitting of a narrow saddle, as is invariably the case with so many modern racing machines, was ill advised in William Palmer's time, "it isn't comfortable and it will probably cause you grievous bodily injury."

At the turn of the twentieth century, many of the Lakes and Dales roads were narrow and rough; they had not yet been made up to the tar macadamed state that we take for granted today. Travelling along such a rough road, where every turn of the pedals jarred the body, gave credence to the early nickname for the bicycle, of "boneshaker." But if the roads were devoid of the surface to give William Palmer a relatively smooth ride, then he did not have to contend with the modern hazard of fast moving, life threatening traffic, with their choking emission of petrol and diesel fumes. Motor cars were still something of a novelty and the prerogative of the wealthy; on the country roads about Kendal William Palmer was more likely to meet only an occasional vehicle in the course of a day's travel. The roads and bye ways that he used were more likely to be shared with horse drawn vehicles or flocks of sheep, than with the internal combustion engine.

With the increased freedom that his cycle brought, William Palmer was able to incorporate a great mileage into his day's explorations. In just the same way as he built up his stamina to enable him to cover over eighty miles in a full day of walking, so he trained himself to be able to incorporate miles of cycling into a day's programme of walking or climbing activities. One of his favourite activities was what he called, "pass storming", which involved a combination of cycling and walking over some of the Lake District's roughest terrain, where his route took him from his Kendal home to follow the round of four severe Lakeland passes. These were Wrynose, Hardknott, Styhead, Esk Hause, from where he made the return journey home by way of Rossett Ghyll and Langdale.

William Palmer was fit enough to be able to comfortably complete this round in a day. One recommendation that he made when travelling with a cycle over the sort of rough ground that the mountain passes presented, "it is quite an excellent idea to remove the pedals from the cycle so that they are not damaged against the many boulders that are struck on the way." He also indicated that, "Before 1914 the storming of these high lakeland crossings, in a day, was considered a pretty good test for a cyclist."

After the end of the war, in 1918 when more resources of both man power and materials were available for the repair and maintenance of roads, the conditions in the Hardknott and Wrynose areas showed a slight improvement. The local authority, the Forestry Commission and the different mining companies that had workings in that part of the Lake District, all set about enhancing small sections of the road system between Langdale and Eskdale. Their objectives were to make access easier for their own vehicles, but the improvements were of such a spasmodic and uncoordinated nature, that the resulting work was hardly of any great benefit to other road users. The road was good in parts, but in others, the rough loose surface persisted as a hazard to both cyclist and his machine. There was no continuous surfaced road for a cyclist to follow and no complete improvement of the road surface was made until after the second world war when an extensive programme of repair was made; not as a result of enemy action, but through self inflicted wounds.

During that period, the Wrynose and Hardknott areas had been used extensively for war time tank training. The rough terrain with its mixture of rock and bog; steep gradients, and enclosing mountain sides presented a challenging environment to both men and machines. The area was a good choice as a secret training ground, for it was remote and difficult of access; it was well away from the eyes of any interested observers, yet relatively close to the armaments factory at Barrow. The approach roads were easily monitored to control the secrecy of training activities, and many of them could be closed without presenting much inconvenience to local people. The combination of these factors made the Hardknott and Wrynose areas an ideal location for men and machines to test their capabilities in difficult conditions.

But the valley between the summits of the two passes and what passed as a road, paid a price. The gouging and churning of the ground that was inflicted by the steel tracked, heavy vehicles rendered the road as unsuitable for other vehicles. As a result of this war time damage, a complete repair of the road system was essential before it could be returned to peaceful and civilian use.

It needs some imagination to envisage what the surface of these two passes must have been like in William Palmer's day. The steep gradients and hairpin bends that present the modern day motorist with such a challenge, were there, but their negotiation was made the more difficult by a top layer of rough scree, that periodically swept down from the close proximity of fell sides during many a winter storm. Stones and rocks of many and various sizes lay uneasily together to form a surface that rattled the rider into a teeth chattering, uncomfortable passage.

William Palmer lightly dismissed the conditions encountered on such rough ground in a matter of fact fashion, by stating that to manoeuvre his cycle, "involved some hard lifting among loose rocks." It has to be remembered that the cycle that called for "hard lifting" was a heavy cumbersome machine that was in fashion before the introduction of carbon fibre frames, and the advent of the multi geared mountain bikes.

To negotiate any of the Lakeland passes, a hard rider faced a number of what appeared to be, interminable uphill slogs with his unyielding steel machine. Once at the top there was at least the consolation of the descent being easier on the lungs, if not more perilous to the limbs, before the steadier, and more comfortable progress in the valley was reached.

Some of the descents that William Palmer faced were hazardous in the extreme. Rossett Ghyll was one of the most notorious. He recorded that however many times he negotiated the sliding rock of the steep descent, it never seemed to get any easier. This descent came at the end of his round of the four passes and concluded the long, lung bursting push up the stones of Sty Head Pass and a further ascent to Esk Hause.

To most car bound travellers of today that enjoy the magnificence of the Lake District's spectacular mountain scenery in warmth and comfort, Palmer's exploits may be regarded as self punishment of the most extreme kind; yet, he did it for his own pleasure. On many of these challenging cycle rides, he was accompanied by a friend. For one reason, he argued that it was much easier to manipulate the passage of a heavy and cumbersome roadster bike up the steep roughness Sty Head Pass if there were two people to share the effort. His method was for both men to get one machine part way, then come back for the other. In this way, the machines were leap frogged up the pass. Two riders could make an easier job of getting their bikes to the top than a solo rider could.

Grunting, muttering, heaving, shoving; in this way the unwieldy machines were hauled up many a mountain pass. "The cycle from being a patient friend and then a neutral, becomes a stubborn enemy, who must be watched all the time." Punctures were inevitable in these conditions and their repair was a time consuming, but necessary business.

By the time the lighter cycles were on the market at an affordable price, William Palmer's interest in the sport for its own sake, had become somewhat dimmed. He exchanged the shoving and hauling of a heavy machine over the rough roads, through bog, slime and mud, for his new found enthusiasm for rock climbing. The cycle again became purely a means to an end as his interest was "attracted by mighty rock towers, great slabs, and deep, steep gullies which called me from afar." But if the flame of enthusiasm for cycling waned and flickered, for a period of about sixteen years, it certainly did not die, although during those years, it played second fiddle to his active rock climbing that spanned the years between 1902 and 1918.

Many years later, in 1944, in the introduction to his book "The River Mersey", he put his interest in cycling in the perspective as he saw it. "I was always more a rambler and a nature lover than cyclist, and the number of miles per day, after high moorland tracks had been reached never mattered."

In the search for material that he could use as a journalist, William Palmer extended his cycling from day excursions in his own immediate area, to venture further by touring through other parts of England, and lowland Scotland. This gave him the opportunity to travel on more level and hospitable roads than the mountain passes of the Lake District provided. In these circumstances, he was quite happy with his own company and preferred to keep moving at a leisurely pace to observe what was about him. He stopped only for food, or to investigate places of interest about which he would write, and philosophically penned the words, "I am not one of those envied persons who can take a wayside sleep in the middle hours of a riding day, but otherwise - where's the hurry?"

He also wrote, "Cycle alone - and no one can correct or refute the impressions which may vary with the company; nor can one be caught out when side-stepping the truth in answering some question." He admitted that when he was cycling with a companion, he never seemed to get into the adventurous or nerve wracking situations that he seemed to encounter when travelling alone. Is there a suspicion that his accounts of some of his exploits might have been enhanced, exaggerated or enlarged to make them just that little bit more dramatic and exciting?

As a journalist, he was ever conscious of making his stories and articles of great appeal and interest to his intended readers The tabloid press of today indicates that news of misadventure or disaster catches the eye and interest of readers more readily than straight forward good news stories. The recording by William Palmer of his walking, climbing and cycling exploits was the very stuff of his livelihood. Who could blame him if he enlivened what may well have been a drab and dismal experience into something that smacked of danger and excitement. After all, he had an editor to please who was always on the look out for something out of the ordinary.

Palmer was the man to supply it. For over fifty years William Palmer regularly satisfied a wide readership with tales of excitement, interest and adventure, that had enough authenticity to make them credible.

Yet, although he cycled alone for quite lengthy periods, he was not regarded by the rest of his fellow cyclists as one who preferred his own company. Social riding was very much part of his cycling scene, and many of his rides were taken in the company of a few friends or members of the local branch of the Cyclists Touring Club of which he was a long term member. He enjoyed and appreciated the camaraderie and companionship; the humorous banter; the kindliness and loyalty of friendship. No wonder he described the cycling fraternity as the very salt of the earth. The CTC which still exists today and has branches throughout the country, is a national organisation that promotes and encourages a wide level of interest in the sport. The club logo of the "winged wheel", was a welcome sight on many occasions to William Palmer when he was touring. It was an indication that the many guest houses or hotels that displayed the sign, provided accommodation that had been approved by the club. He was ensured then, as are many riders today, that there was a warm and comfortable welcome, at the end of a hard days ride. "I am not finicky about quarters, or whether there is another bed in a large room, but I do like good food, simple but well cooked and cleanly served."

In later life, he tried to visit, whenever possible, the annual northern rally of the Cyclists' Touring Club which was often held in the Bowland area of Lancashire. At this camping weekend, up to 300 tents of various sizes were pitched in close proximity to each other. Their occupants "preferring friendship to comfort." He wrote, "My wife commented that the walk through the camp was a progress; I called it a pilgrimage. There were so many different tents we had to visit. Campers had come from all parts of Lancashire, Cheshire, Yorkshire and Derbyshire whom we knew. I admit that I never met so many cheery and active lads and lasses, with elder riders too. Most we had met on distant roads and camp sites; the rest were familiar callers."

Each year, until he was well into middle age, William Palmer spent days away from home, pedalling his way through what he came to call the "Verge" country; the approach lands to areas of greater, and more obvious interest. He planned his cycle routes of exploration in much the same way as when he was travelling on foot. He plotted out his course, and made notes of interesting places to visit, in anticipation of writing articles about them that would be suitable for publication. In this way, cycling gave him another open door to the market for his freelance articles; he was able to submit pieces to those magazines that were included in the cycling press.

His travels by cycle took him throughout England and Scotland, although when his journeys were of a considerable distance, he often made use of rail transport either to take him to his starting point, or bring him back from his destination, so that he only cycled one way.

William Palmer (front left) at the CTC Rally,
Bolton by Bowland, 8th May 1932.

Another form of tour cycling that he enjoyed was what he described as "looping." This involved riding up one side of a river, crossing by a bridge to return down the other side to complete a loop. He linked these loops together to develop a chain of discovery that lasted all the days that he needed to explore a river system.

When William Palmer ventured on these far ranging travels, he wasn't always content to stay within the security of the frequented and recognised road network. On many occasions, he sought out wild and lonely places to test both himself and his machine over difficult ground. Once a full reconnaissance had been made over the whole area, and the route plotted, ridden and recorded, he was then in a position to write about it for a cycling magazine, in order to persuade other cyclists to take up the challenge.

In cycling terms, he described himself as a "trouble seeker" not because he put himself at variance with the law; this was to the contrary for he was always careful to make sure his machine was well maintained. It had to be, to survive some of the punishment to which it was subjected. William Palmer's "trouble seeking" meant looking for the really difficult terrain across which he could plot a route to test the "hard men" of the sport. It often incorporated much uphill walking and many of these tours took place in the Yorkshire Dales where the crossing of bog infested heather moors were incorporated into his routes. They were almost guaranteed to provide him with some "excitement."

One of his worst experiences of this type of rough riding came when he travelled south from the Tan Hill Inn, which is reputed to be the highest pub in England; he rode the Pennine Way, before the ramblers made it their own. William's journey took him across the bleak and deserted moorland that lies to the south of the isolated inn, to Keld in Swaledale. That lonely crossing was done after a period of heavy rain and when later William Palmer recorded the details of the journey, he began his account with, "I looked forward to adventure. And I got it." What he got in the way of adventure was expanses of deep green moss, the brilliance of which is known to all walkers as an indication of places to avoid. There were lagoons of water that hid the security of a route; rushing streams that had to be crossed, and soft, peat hags that sucked both man and machine into their black yielding troughs. "The floor was slippery, and now and again I was stopped, leaning against the cycle, at other times the cycle was leaning against me."

There were only infrequent occasions on this particular journey, when he was able to ride his machine over stony surfaced outcrops, before being faced with more problems of a wet and glutinous nature. Even today, this particular moorland crossing is one of the most uncomfortable sections to be negotiated during the course of the now popular long distance walking route that is known as the Pennine Way. It's crossing is anticipated with trepidation; to a walker with a heavy backpack it presents sufficient discomfort in wet conditions, let alone the problems faced with the addition of the cumbersome manoeuvrability of a recalcitrant bicycle.

William Palmer advocated that travelling by cycle was one of the best ways of getting at close quarters to animals and birds, whether it was when travelling across rough northern moorland, or making a leisurely journey through leafy southern lanes. He wrote, "In many rides I have had fair fortune in seeing 'sporting creatures'." Cycling has the great advantage over other forms of travel in that it gives an almost silent approach along highway or byway, and enables the rider to quickly stop and observe any wildlife incidents encountered on the way. These are factors that are envied by a nature loving motorist. Cocooned in a shell of fast moving metal and glass, the car bound naturalist is handicapped by a noisy engine and the wastage of time spent drawing to a halt and immobilising the vehicle, before any observations can take place, by which time, the "nature" has fled. The magic of the moment has been lost. But not so for the cyclist who is more in harmony with the nature of the countryside.

It was on one of his lonely cycle rides within not too many miles of his Kendal home that he recalled a most memorable evening. "The moist clouds split, and through narrow gate came golden sunshine. And in the ray of glory, with the light touching its wings, hung a great buzzard, big as a minor eagle."

Although most of William Palmer's cycling was done in the northern counties, he travelled extensively in Scotland as well as Wales which was within easy reach of his Liverpool home. Sometimes he ventured into the southern counties, and had cycle tours in the south west peninsula where he explored Devon and Cornwall. It was on one of these tours that "one may happen on adventure, when and where it is least expected." This particular adventure took the form of a verbal assault, accompanied by a badly aimed rolling pin from an irate mother, because William Palmer had stopped to ask directions of her child. "I told him that I wouldn't have him talk with strangers," she said. Sixty years later, children are still being given the same sort of advice.

His lengthy cycle tours took him to remote hamlets and villages, where he found that there was a great advantage in travelling alone. The sight of a lone rider with only his cycle as a companion, was an unusual enough occurrence to evoke the interest of local people in the bar of a village inn. When William Palmer entered a pub in a strange village in search of refreshment, he was quick to assimilate the mood of the place; a nod here, a quiet word there, as he ordered a soft drink to take away to a quiet seat. William Palmer was not a man to intrude on any conversation that was already taking place, but he was a good listener; he was prepared to bide his time, and when he judged the moment to be right, he threw in a chance remark that conveyed his interest.

Like an angler casting a fly to lure a curious fish, he caught their attention; he had them hooked. The people talked in response to his words of encouragement and genuine interest. They enjoyed, as do most folk, the opportunity to talk about the happenings and highlights of their lives. William Palmer listened, he remembered what they told him, and as soon as he was able, he recorded the tales which were often sent off home in letters to his wife.

Later, when he presented these records in the form of articles for magazines, or chapters for his books, he acknowledged that if the local folk had realised that the true purpose of his cycling tours was to gather material for publication, they would perhaps have been reluctant to talk so freely.

When William Palmer moved to Liverpool from his Kendal home in 1918, he was already a well experienced cyclist, who had covered literally tens of thousands of miles in all sorts of weather, and over all sorts of road conditions. His membership of the Cyclists' Touring Club had established friendships with many fellow enthusiasts in different parts of the country. He brought this experience and personal enthusiasm for cycling with him to Liverpool, where he was instrumental in developing a local branch of the CTC in the city; a branch which is still flourishing today.

The success of any good club is dependent upon those activities which are organised for the benefit of the members to enable as many as possible to take part. With the practical cycling experience that William Palmer had gained, together with his organisational skills and his route finding capabilities, it was inevitable that he was involved in organising the programme of events. He regularly joined in the weekend and mid week rides into North Wales and the Cheshire countryside that he grew to know so well. He experienced the feeling of well being that hard physical exercise can bring, and appreciated the camaraderie of the communal chatter over a cup of tea at the end of the day.

As an experienced cyclist, and an established "hard rider", he was very much aware of the responsibility that the older members had for the youngsters in the club. He tried to inculcate into the Mersey branch of the CTC, the philosophy of the Fell and Rock Climbing Club, whereby on occasions, established members were prepared to forego their own satisfaction of a day's climbing, to devote time to encourage younger members to develop in the sport. His mind went back to his early rock climbing days, when over endowed with enthusiasm and self - confidence, but lacking the necessary rudimentary safety skills and equipment, he tried to emulate what more proficient climbers could do.

William Palmer felt that in much the same way, the young cyclists of Merseyside wanted to show that they too could be "hard men", and keep up with the top riders when out on club rides. They tried to keep the pace going over the route, to demonstrate that they too were able to cycle through difficult and adverse weather conditions, just as well as the more experienced riders.

William Palmer was well aware that this demonstration, of what would now be regarded as a "macho" image, was not always in their best interests. Some of the younger riders were physically immature, and they ran the risk of injury if their youthful enthusiasm was not curbed. This was especially so in the cold and chill of winter months, when their lack of protective clothing and equipment, allied with their inadequate experience placed on them great physical demands and discomfort. William Palmer realised that this could result in them eventually giving up the sport.

He advocated a policy of nursing the youngsters along, as the young rock climbers of Wasdale, Langdale and Coniston had been encouraged to develop their skills. "Mek haste slowly", was an old Lakeland saying that William Palmer adopted. This point of view was not always met with the full approval of the rest of the experienced members of the club. They regarded his suggestion that the inclusion in the season's activities of a number of shorter rides, was a "watering down" of the Club programme. They preferred a full complement of long day rides in the open countryside, and some even saw the inclusion of shorter rides as a threat to their own personal enjoyment of the Club's activities.

But William Palmer was a determined man and he persisted with his proposals in which his reasoning was sound. The older and established members of the Club were well equipped, both in machines and protective foul weather clothing. They had, over the years built up their physical, and material resources along with their finances and experience. They were able to come through the harshness of a long , cold ride if somewhat wet and a little uncomfortable, knowing that at the end of the day, they had the reassurance of a hot bath, warm clothes and the comfort of their own homes to look forward to. Many of the young lads in the Club, that William Palmer was anxious to encourage, were raw novices to the sport. They were poor in equipment and poor in pocket. Some were shipyard workers in the great docks of Liverpool, Wallasey and

Birkenhead that once lined the River Mersey with an almost ceaseless bustle of industry. Many lived in lodgings away from home and were not always kindly received by their landladies as they returned wet, muddy and cold, from long hard rides.

To keep the interests of these young members, William Palmer also suggested that the club programmes should include plenty of indoor social activities, following the pattern of the Camping Club of Great Britain and Northern Ireland, so that when bad weather curtailed their cycling, the members could still enjoy the companionship of a Club night, as memories and experiences of former outings were shared. He persisted with his arguments, his ideas were adopted, and the club flourished.

William Palmer maintained his interest in cycling for most of his active life and although he eventually owned a car, and would use motor transport when it became freely available, he never bothered to learn to drive. His daughter acted as his chauffeuse. He only gave up his long distance cycling when he was well into his sixties.

In the preface to the 1948 edition of "Odd Corners in English Lakeland", when he was aged 70, he wrote, "I have ridden many wild tracks, but my cycling days personally must be regretted as over." Although he was then unable, physically, to involve himself in the practice of the sport, he still obtained great pleasure in planning routes for others to follow.

"When the maps come down on the table, my wife protests; There you go - putting folks up to rides you don't go yourself.' He admitted that his wife's objection was sound. "It's easier to plan than carry out a continuous girdle of the Lake District. I've done it all: in winter and in summer, but never at a continuous trip."

William Palmer's planned cycle route, which he called "Girdling the Lake District", called for stamina and determination on the part of the rider, and reliability, on the part of the machine.

The route that he planned incorporated a visit to each of the 13 lakes, and he expected the rider to be able to become a walker, in order to carry out the crossing of steep mountain passes.

"Such a girdle route is worth the while of any adventurous youth or pair of youths." To gain the maximum amount of enjoyment from such a ride, William Palmer promised them, "plenty of collar work and a lot to see that is hidden from the summer tourist." He knew only too well, the difficulties that rough and hilly terrain presented to the cyclist, but he saw no disgrace in pushing the single geared, heavy cycle of his youth up many a steep slope.

The advantages now available from the multi - sprocketted assembly that provides up to twenty options and enables a modern rider to pedal up severe inclines, would not have seemed possible to cyclist of Palmer's day. He wrote, "Hill climbing on a cycle needs skill and careful use of power rather than brute strength." He would be surprised at the ease with which modern mountain bikers of even mediocre skill, negotiate the roughness of Black Sail Pass, between Ennerdale and Wasdale; or the way in which multi geared touring lightweights forge to the top of Honister Pass. William Palmer recalled some hard climbing on the unmade surface of that pass, which leads from

Buttermere to the Borrowdale valley, when he wrote, "There is stiff pushing indeed for a short distance which seems to be long. A strong and patient person will win through, but anyone less blessed will be sorry."

In the late 1940's when his book "Wanderings in Surrey" was published, he wrote.

"My esteem for cyclists is well known. My riding for years was in the wildest country and under rough weather conditions, and all their experiences have been shared. My simile, - the salt of the earth - applies more to those who sweat the long leagues through heat and dust and glare, who ride by night as well as by day, and whom it takes a rough storm to beat off the road once a start has been made." This representation of the salt of the earth cyclist could have been a direct description of William Palmer himself.

Chapter 7 - THE COUNTRYMAN

When William Palmer grew up in the Lakeland village of Bowstone during the latter part of the nineteenth century, there was no readily available, consumer based entertainment for him to look forward to during the course of an evening or weekend of relaxation. Once he was away from the restrictions of the classroom or work place, he joined in the home made entertainment with other lads, that provided a lively alternative to the humdrum routine of day to day life.

As with many other villages, the turn of the century saw the emergence of local football, rugby, and cricket teams, that created a focal point to the sporting life of a community, but William Palmer preferred to find his own amusement and pleasure in the open countryside, or along the river bank. While he enjoyed the company of others, he was in no way a team man, and this was reflected in the sports that he pursued where the success and enjoyment was as a result of his own individual efforts.

The village school that he attended in Burneside with its academically oriented curriculum afforded no opportunity for youngsters to become involved in competitive team games. In common with other Board Schools, the sporting activities that form part of today's curriculum were unknown. Parents paid the "school pence" for their children to learn, and not to play. There were no school football or rugby teams to stimulate physical activity and encourage youthful enthusiasm. All that was available to the children in the guise of physical education was the ritualistic form of "drill." Four straight lines of children spread across a school yard. At the brisk command of "Arms' length, feet apart", they shuffled within their regimented rows to meet the required distance. Then, with evidence of intent, if not with execution, they jumped; "Astride together, astride together." Their practised movements clattered, heavy booted and expectantly in unison, to the teacher's commands.

But away from the monotonous activities of the school yard, with the whistled or barked commands, there were fathers and brothers, uncles and cousins, as well as men of the village with more exciting sport on offer. These elders were prepared to kindle youthful enthusiasm by spending time with a keen and interested lad. They shared with him their knowledge of the countryside and showed him the finer points of, what are now referred to as, country pursuits. Hunting, shooting, fishing are seen by many today as the prerogative of the privileged. In William Palmer's youth, they were an essential part of the country life style of the ordinary folk. When money was scarce in a household, and there was little food in the pantry, a lad's success in setting a snare for a rabbit made all the difference between a meal on the table and a family going to bed hungry. In this respect, the Palmer family was no different from all the rest. There was food for free in the fields and hedgerows that surrounded his home, for in the years of William Palmer's youth, the surrounding countryside teemed with wildlife.

There is an extensive network of small streams and becks that contribute their waters to the greater river system of the Kent, from which Kendal takes its name. The pools and deeps held salmon and brown trout for the experienced anglers; while the shallows yielded minnows and sticklebacks to the anglers in the making. Unpolluted meadows and open grassland were the homes of healthy populations of rabbit and hares, while there were numbers of red grouse on the once heathered slopes of the Shap fells. Although these were the creatures that were destined for the chase and the pot; other animals such as fox and otter were hunted for the chase alone. William Palmer enjoyed the thrill of the chase for whatever reason, but he left the humbler and more insignificant creatures unmolested. William watched, observed and studied these, and this enabled him to build up a rich background of countryside knowledge which was to stand him in good stead in later life.

The house in which he was born, Number 4, Old Bowston, or Boustone as it was once written, backed on to the fast flowing, River Kent. He described his birthplace as "a hamlet in the county of sparkling trout streams." That same river, when seen in the right conditions, still sparkles today as brightly as it ever did. It dances over shallow steps of the water worn stones of its bed in carefree splashing leaps, catching the flashes of light as sunshine filters down through overhanging branches.

It was almost inevitable that with the sound of the splashing beck in his ears, almost from the moment he was born, the young Palmer would grow up with a fascination for the river. As the river became his playground, his interest deepened in the wildlife that was dependent upon it, and he became keenly involved in all its related sports.

Fishing was high on the interest list of most of the boys in the village, and William started his angling with the most simple of home made tackle. There was plenty of natural raw material from which the lads made their own rods, for springy ash stems grew in profusion along the river bank. These were easy to cut, and easier to find in an area rich in coppice woodland, the remains of which still erupt on the eastern bank of the river opposite William Palmer's former home. It was also possible to obtain suitable rods from a hedgerow, where the smooth grey stems of ash, with their black imprinted lengths of miniature horse shoes, made fine tackle.

There was no substitute freely available for the fishing line, so the river banks were eagerly searched to recover odd lengths of tangled discards from among the riverside debris. The lads begged other pieces of line from the men in the village, who had long since advanced to the use of commercial and more sophisticated equipment.

Setting up the tackle for a day's fishing was a simple matter for the youthful William Palmer. A scrap of line, when fastened tightly round the tip of the "rod", was adequate enough to dangle in a pool without the sophisticated luxury of a reel. A wriggling bunch of worms was ideal bait, for when freshly dug and securely fastened to a hook, this proved an irresistible lure to many a lurking trout. Although the young Palmer attempted to make his own hooks by using pieces of bent wire, these were not really adequate. He found it virtually impossible to hone the wire to the necessary degree of sharpness, in order to ensure immediate penetration when the fish took the bait. He found it equally difficult to shape the barb which was necessary to prevent a hook from coming loose, when a successful strike was made. Catching a trout was difficult enough, without losing it at the last minute through inadequate equipment. For this most essential part of his fishing tackle, William Palmer had no alternative but to make the trek into Kendal to visit one of the specialist fish hook manufacturers. He was able to obtain the hooks in a variety of sizes, but as they cost a penny each for the most useful size, this ensured that youthful anglers took care not to lose their tackle. William had no such difficulty or expense in obtaining bait, for there was an abundance of freshly dug worms that were used to tempt the silver prey. It was on tackle such as this, that William Palmer caught his first trout.

When he went fishing as a young lad, he did not have the bother of completing the requisite paperwork to obtain licenses or permits; he had no fees to pay to a water board or a river authority. Fishing in William Palmer's youth was a carefree, spontaneous pastime. There was sport in the Kent and other Lakeland rivers, lakes and tarns, for all to enjoy, as and when they pleased.

If the home made rods of the type used by William Palmer and his pals, did not have the strength or suppleness of the spliced cane rods that were commercially available, they had the greater advantage of being easily replaced when they were no longer serviceable. They proved to be quite adequate for his fishing expeditions until the time that he started to earn a wage and had some money of his own to spend.

The first manufactured rod that William Palmer bought cost him, not only a hard earned shilling, but also his mother's displeasure for supposedly wasting his money. This rod became a bone of contention between the two of them, for his mother was a woman who was used to maximising every penny on, what she regarded as, essentials. She saw no sense whatever, in what was in her opinion, the squandering of a shilling on a fishing rod, and made sure, in no uncertain manner that William was well aware of her feelings on the matter. To avoid any further recurring bouts of maternal anger over his purchase, William kept his newly acquired tackle well hidden in the blacksmith's shop. Out of sight, out of mind, was his policy and this resulted in fewer cross words between them.

Much to his delight and his mother's disbelief, the contentious rod paid for itself with the first good sized fish that William caught with it. This occurred on a Good Friday, when he was fishing one of the deep, dark pools of the Kent. He had just succeeded in landing his fish and was still trembling with the excitement of the moment, when the parson's wife stopped him. She offered him a shilling for the fresh catch, for although its tail had flapped its last, the scales still gleamed with colour. There was no hesitation in accepting the good woman's offer. To an impoverished youngster, a shilling was more important than the visual evidence of his achievement; he was glad to take the money. An appreciation of its true value is realised when the price paid for the fish was equated against the cost of William's fishing rod. In comparison with present day costs, it seemed a somewhat expensive fish. William may have been fortunate in the deal as the parson's wife had made no provision for her husband's Good Friday dinner, and this may have been part of the reason why she was prepared to pay such an escalated price for the fish; to satisfy not only her husband's appetite, but also his religious principles. William eagerly accepted the coin, and hoped that he had not lost out on his estimated weight of the fish. He decided against accompanying the woman home to weigh it correctly in case it should be found in her favour.

William Palmer progressed from the boyhood technique of dangling a worm into the beck, to the more skilful sport of fly fishing. He learned the craft of fly tying from men in the village and was equally happy casting his fly on lake or river. It was the night time angling expeditions that left him with lasting memories of good sport, when even though the long hours of darkness spent with a companion rowing up and down a lake, may not always have resulted in a good catch. But memories of drifting in a boat, on a calm lake, where the reflected mountains were mirrored in the moon silvered water, was ample compensation for an empty basket.

He was not averse to a spot of poaching, especially when the salmon were running upstream, and his leisure time was limited for legitimate angling. Fishing is a marvellous sport for passing the time away, but not so satisfactory if a fish is required in a hurry. His learning years of watching the river and noting the movement of the fish at appropriate seasons, gave William the knowledge of where the best specimens were to be found.

Some of the favourite pools were deep and partially hidden under an overhanging bank. It was in these dark, quiet places that the big fish were to be found. They lay at an angle to the current, tucked in with head to the bank and tail moving slightly in the wash of the stream. Pounds of succulent salmon were lying there for the taking.

One illegal method by which William caught salmon was with the use of a fish spear. In a "History of Cumberland [1793 - 6]", William Hutchinson described the use of this implement in an early form of fishing that took place in the shallow waters of the River Derwent at Workington. The method was for a fisherman to ride, on horseback, into a shallow stony area of the river. The salmon were bound to reveal themselves as they forced their way upstream to the spawning grounds. It was then a simple matter for the rider to thrust a spear, lance or leister, as the implement was sometimes called, into a struggling fish. When William Palmer used a similar instrument, he didn't resort to the use of a horse. He was content to lie on a bank, or stand motionless in the stream to wait for the appearance of a suitable fish.

In Burneside, there was a blacksmith who was obliging enough to hammer out a three pronged spear and weld it to a shaft. Thus armed, the young poacher, who was usually accompanied by a pal, worked the pools of the river by the wavering light of a lamp. Although this was an illegal, method of fishing, its successful execution still called for skill and judgement; for co-ordination of hand and eye was essential to make a successful thrust, and defeat the refractory properties of the water.

On a few occasions, the voices of the beck watchers or keepers interrupted their fishing, but when this occurred, the light was quickly doused, the fish already bagged and the spear were hidden, to be reclaimed at a more opportune moment. The spear that William used was a fore runner of the illegal gaff that is much used by modern poachers.

Another illegal method of catching fish, with which William was familiar, was the use of a "lath" or "otter board" as it is sometimes called. He found the best place to use this equipment was in the high mountain tarns, where there was often a good stock of brown trout, and where the remote locations afforded less chance of the poacher being apprehended. The "lath" consisted of a piece of board, to which lead was fastened on one long edge. This ensured that the board floated vertically in the water. A number of lines were fastened to the board, and the hooks were baited with worm, mayfly or beetle.

The poacher guided the board by a long line which was attached to the "lath." This enabled the board to be controlled after it was sailed out towards the centre of the tarn. The poacher tweaked the board into position so that it drifted with the wind, over likely waters that held the unsuspecting fish. William's resulting catch was usually more than adequate to provide a supper of "poached trout", and was ample compensation for carrying the weighty board many a mile.

As he matured, William Palmer's fishing became dominated by the respectability of the rod and line man, but he still enjoyed the excitement of joining forces with a local poacher to work a beck.

Anyone who has tasted the thrill of an illegal fishing trip can relive the excitement conveyed through Palmer's own words, as he described his experiences in the beck. There is the waiting and tension in the inky blackness of the night; the expectancy of reward as a bump is felt in the net before the slippery, wriggling body is thrust ashore. Above all, there is the stomach churning feeling of apprehension as approaching voices with their accompanying flashing lights, raise the sickening possibility of being caught. The agony of a breathless, heart pounding chase with the anticipation of a clutching hand or overhead shot. These were all experienced by William Palmer before he reached the sanctuary of his home. For those of a law abiding nature, who have not experienced such events, let William Palmer tell the tale, as he tried to avoid the bailiffs.

"There was no time for contemplating the darkling stream, or for shivering on the brink - the terror of the police court is mightily great. In the three of us stepped; knee deep the cold was horrible, waist deep the feeling was worse, but before bottom was touched the water was neck high and the chill seemed to freeze our very marrow. The poacher, we still had confidence in, for he had been in scores of similar tight corners; with arms outstretched, he pressed us to the bank. I heard them moving high above our heads, and saw the gleam of lanterns. The chill of the water was forgotten in that breathless five minutes."

The following morning, there was salmon for breakfast for William Palmer.

After he moved away to Liverpool to further his journalistic and writing career, he occasionally returned to his former village to visit members of his family. To his great regret, he found that the freedom of his boyhood river fishing on the Kent, was no longer available. Landowners and syndicates had claimed their stretches of water with their own fishing rights. Strangers had taken over the lively beats. The fun had gone.

Another of the country sports with which William Palmer was involved during his youthful years, was hunting. The nearest pack of fox hounds was kennelled at Ullswater, but the village lads and men also followed the pack of harriers that were housed on the nearby Potter Fell.

The pursuit of the hare is a much older sport in the Lake District than fox hunting, for it was practised long before packs of hounds were organised to seek out the "red vermin." Evidence of this can be traced on the sign boards of old time inns, where "The Hare and Hounds" was once a common name of a hostelry and was used as the meeting place of the hunt. When William went out hunting the hare, the day long chases in which

Ullswater pack meet for a day's hunting.

Tommy Dobson (left) with Willie Porter - Eskdale.

he was involved, roamed over open fell and farmland, and usually resulted in a few of the participants bringing something home for the pot, which was a welcome addition to the family larder.

Fox hunting in the Lake District, is not as old a traditional sport as is commonly thought. The packs of hounds that today are specially bred and trained to work the rough fell country have only been established for about 150 years or so. Prior to that, it was a motley collection of terriers and farm working dogs that were assembled to accompany groups of farmers and villagers to seek out and rid themselves of foxes, which were a major threat to their stock at lambing time. By the time William Palmer was old enough to follow the hunt, at the age of about ten, some of today's recognised Lakeland packs of hounds had already been established.

A near neighbour was Tommy Dobson, who was a former bobbin turner from the neighbouring village of Staveley. "Laal Tommy" as he was known, because of his diminutive size, became one of the Lake District's most famous huntsmen, through the success that he achieved with the pack that he himself created, the Eskdale Foxhounds.

William joined Tommy Dobson on many a hunt in the rough high ground of Great Gable, Pillar, and Scafell areas, and he recorded that he was with Tommy on Esk Hause, when on one occasion the hounds ran a fox almost at their feet. There it was killed instantaneously. This was one of the last hunts that Tommy led his hounds, before he retired as huntsman. The descendants of Tommy's pack continue to hunt today, but do so under the name of the Eskdale and Ennerdale Foxhounds.

When Tommy died in 1910, while staying in Little Langdale, William Palmer was a member of the small funeral cortege that followed the huntsman's coffin from Langdale to Eskdale. This small procession, which was led by a horse and cart that acted as a hearse, left Little Langdale in gale force winds and rain, to cross the two passes of Wrynose and Hardknott. The journey was completed with only five minutes to spare before the appointed burial time at St. Catherine's Church in Eskdale. The wind, rain, and the hospitality of a wayside inn had caused some considerable delay to the cortege .The grave now has an impressive headstone, which was bought as a result of subscription among the farming community, and this marks the huntsman's final resting place.

The pack with which William did most of his hunting was the Ullswater foxhounds. He was never a fully paid up member of the hunt, but he enjoyed the opportunity of following the hounds whenever he could. If William Palmer was walking the fells and heard the blood stirring cries of hounds, or the insistent summoning of a horn that told him a hunt was in progress, he needed no second bidding to change his route. Fell walking was readily abandoned for the chance to follow the hounds.

The Ullswater pack was formed in 1873, only a few years before William Palmer was born, and its most famous huntsman was the legendary Joe Bowman, who was known locally as "Auld Hunty." Joe died at Glenridding in 1940 at the grand old age of 90. He was reputed to have killed more foxes during his hunting career than the better known John Peel of Caldbeck, whose hounds are said to be the ancestors of the modern Blencathra pack. The Ullswater pack still hunts over what could be termed William Palmer's home ground. They cover the Kentmere and High Street Fells, and extend over to the Fairfield and Helvellyn ranges. This was all territory with which William Palmer was familiar.

In the Lake District, hunting now takes place between October and March. The hounds are followed on foot, or by car, but usually it is only the huntsmen and a few fit enthusiasts that persevere to the high ground. Nowadays, most folk are content to take up, what they hope will be, a good vantage point in a valley and scan the fell sides with binoculars to follow the action taking place. Others take advantage of communication technology by using two way radios between the man on the fell and the cars in the valley in an attempt to follow the movements of the fox and the chasing hounds to its ultimate conclusion.

William Palmer was no valley man. When he went out with the hunt, he followed them to the high ground. He kept as close as possible to the hounds that nosed their way among the splintered rocks where the "redskins" had the advantage of home territory. He followed the hounds for hours on end, sometimes over the snow and frost - covered fells of wintertime, and this often involved getting into difficult situations. On one occasion when he was well to the fore, he had to be pushed up a narrow gully by a hefty shepherd, when the party were thwarted by icy snow that had frozen on to smooth rock. Their dominant role of the men as the hunters was reversed, for the hostile conditions in which they found themselves, put them in a position of some danger. While the nimble fox was able to negotiate the difficult terrain, some of the hounds and men were not so fortunate. On that particular hunt, three of the hounds fell to their deaths, and the hunt was finally abandoned when one of the followers slipped and had to be rescued.

When, as often happened, an excited hound followed a scent to the point of becoming cragfast, their rescue was described by William Palmer as, "This work is both exciting and dangerous." The rescuers had to scramble over rock, or in some cases be lowered to the ledge on which the animal was trapped, by a rope which was carefully handled. It required great skill, and courage on the part of the rescuer to get as close as possible to the trapped animal. Once on the ledge, there was then the problem of actually recovering the hound. "Of course a hound has to be carefully handled under such circumstances; an error in holding it may cause it to writhe free of its carrier to certain death on the rocks below." William Palmer, in his book "The English Lakes" [Kitbag Edition] claimed, "fox hunting among the fells is a matter of keenness and endurance. It requires a higher state of health and tenacity than ordinary hill walking or scrambling. Moreover the days are short, the hill tracks rendered difficult in darkness, and the day may end easily in mist and storm."

A day's hunting was, and still is, an arduous affair. Many miles are covered by the huntsman, his hounds and a few hardy followers as they traverse the fells and ridges between the different valleys. A fox that is initially followed may give the hounds the slip, and thus cause the dogs to lose the scent of one fox, before picking up that of another. A report of a hunt in "The Whitehaven News " of 1893 gives an indication of what was then involved in hunting. It is typical of the activity in which William Palmer took part, and about which he described in detail in chapters of his book "In Lakeland Dells and Fells."

This particular hunt was led by Dan Tyson, of the Wasdale Head Inn, with his pack that was known as the Scafell Hounds. Dan Tyson had succeeded Will Ritson, not only as landlord of the Inn, but also as the local huntsman. Will had retired from the inn to live in a nearby farmhouse.

"It seemed that Auld Will could not sleep at nights for disturbance caused by the fox; and so impudent had the varmint become, that he actually prowled about the neighbourhood in daylight. Dan and his pack were not long in getting on the scent, the dogs going off at a rattling pace.

After a grand hunt, the fox took earth in Buckbarrow, but the terriers sent in, soon put an end to the sly one's existence."

Two days later, the pack went over Black Sail Pass to Gillerthwaite where night time raids had been made on lambs and poultry. This hunt started at the normal time of five in the morning when the scent of fox was fresh.

"The dogs tore away at great speed towards Hay Cocks, Thewat Tarn, through Matlin Cove, Blackholme by the head of Windy Gap, over White Pike across the Pillar Mountain, and as far as Black Sail, where they put the varmint off, and followed on his track under the Pillar Rock, by White Pike through the coves at the head of Bleaberry Gill. The fox doubled back to Iron Crag opposite Ennerdale Lake to Todhole, the strongest bield in Ennerdale. Here he took ground after giving a chase of twenty three miles, at nine o clock in the morning. Fox hunting in the fells is no joke, and the man that's able to go through a day with Dan Tyson and smoke his pipe before he goes to bed, won't die of rheumatics in his hip joints."

William Palmer was the sort of man who was able to stand up to the rigours of a day's hunting with the likes of Dan Tyson, and although his abstinence from tobacco ensured that he didn't smoke a pipe at the end of the day, he was still able to enjoy the entertainment and refreshment that concluded the day's activities in a local inn. A hearty supper taken in good company, was followed by an enthusiastic, if on some occasions, a less than tuneful rendering of hunting songs. Although limbs were weary after many miles of ups and downs on the fells, the voices had had little exercise, and were in fine fettle after a successful day with the hunt.

Cockfighting was another of the sports with which William Palmer was familiar. There are extracts from many parish records which show that this bloodthirsty activity was common practise in the Lake District for over three hundred years. The fact that the "sport" was once deemed acceptable to society at large, was indicated by the fact that it was practised by clergy and nobility as well as the lowly village folk. It was an acknowledged part of country life. The owner of a winning bird was recognised, feted and applauded for its achievement, even though he had not earned any accolades through any efforts of his own. The subsequent triumphant posturing of victory was not only practised by the fighting cocks!

Cockfighting has been an illegal activity in Britain since 1849, although the banning of the "sport" was an insufficient deterrent when large sums of money could be won if one had a successful bird. Records show that a series of prosecutions took place, and were reported in the Westmorland Gazette, as late as 1938. There are still murmurs of cockfights being held in parts of the county during the 1960's. Indeed, some may still be furtively carried on in secret locations, to satisfy the innate savagery of some followers of the activity

When Palmer was a lad, there were still a number of former and active participants who lived in his village. They were able to describe to him in great and gory detail, some of the monumental contests in which their birds had been involved. This was over forty years since the act which outlawed the "sport" had been passed. Although it was not then the acceptable part of village life that it had once been, youngsters were still given an indoctrination into the barbaric activity. Egged on by their elders, in the name of fun, the lads fashioned make shift spurs out of darning needles, which were fastened to the natural spurs of bantie cocks. These birds were then set to fight against each other, to the mounting excitement of men and lads.

William Palmer was told stories of the fighting cocks of former years, who were nurtured by their owners with as much care as they gave to members of their families; in some cases more so. The proud, strutting cockerels were specially bred and fed on the finest food, in the hope that the genetics inherited from their former successful fighting ancestors would combine in a champion strain. It seemed something of a waste, when their inevitable short term of life was considered, for few birds survived uninjured, to live to fight again.

Before the activity became illegal, Shrove Tuesday was the popular day for cock fighting in many a Lakeland village, where most had their own cock pit. Crowds gathered round the hollow to wager on what they considered to be the best birds. Even in the valley schools, the schoolmaster collected his cock penny from each boy, before leading them off to the cockpit, where he acted as umpire to the contests between the lads' birds. The fighting was fierce and bloody, as most contests either ended in the death of the defeated bird, or an entrant was withdrawn by its owner, when it became disabled by severe injuries. These were caused by the raking of the opponents sharpened metal spurs, which were fastened to the birds' own. Often a bird was so badly injured, it had to be destroyed, and that was not always a humane procedure.

There was an almost sheepish denial of any personal involvement in the so called "sport" by William Palmer, when he wrote about it in his book "Odd Yarns of Lakeland." The knowledge he gained through acquaintance with former "cockers", and what he saw for himself when present at the "mains", [the embracing term for location and holding of fights]; ultimately brought about a realisation of the unacceptability of a pursuit that was cruel in the extreme. When he was in his teens, he inadvertently became involved with secret cock fighting when he strayed into a "mains" while walking in the Black Combe area of Cumberland. This was towards the end of the 19th century when he was aged about seventeen. Somehow, he unknowingly infiltrated the ring of watchers who were standing guard. They held their position just a warning distance away from the cockpit. It took him some little time to become aware of what was going on, but was soon enlightened when he joined the excited crowd of rival groups of miners and iron workers from Millom and Egremont.

He stayed there long enough to watch a number of fights, "where fighting cocks were at their messy game, main after main was concluded as one champion or the other was knocked out with the steel clad spurs of the victor." No doubt, as with most young men, much of the fascination was the intrigue of being involved with the illegal. Eventually, the fights were suddenly interrupted and brought to a halt by cries of alarm which forced the crowd to scatter. The birds were separated from each other, and hastily thrust into sacks, as their owners tried to make their escape. Along with others, William Palmer had to make a speedy retreat over the fells, when large numbers of police appeared in force. They had been tipped off that cock fighting was taking place and were determined to catch those involved.

It was only because William Palmer was agile and had an easy familiarity with fell country, that he was able to avoid capture by the police. "I had to race against a constable of pretty big proportions. I did not hesitate, but made for the buttress by the stream, reached it first by thirty yards and went down at my best speed." He did not stop running until he had crossed the Duddon Bridge, which then marked part of the boundary between Cumberland and Westmorland. Once across the bridge, the Cumbrian police had no jurisdiction to continue their chase, William was safely away. The other participants, many of whom were iron ore miners, were not so fortunate. They were all captured and fined, and most had their birds confiscated.

Many years later, when William Palmer had assumed the mantle of respectability that comes with the responsibility of family life and a steady job, he was reminded of his narrow escape from the law. He was chatting with an old farmer, and when the conversation turned to cockfighting, the realisation became apparent that both had been present at the Black Combe mains. The farmer had been one of the organisers and remembered William's intrusion. He was told that he had only been allowed to stay on at the mains, because of "his extreme youth."

Other "sports" in which William Palmer was involved as a lad were others that also became illegal, those of otter hunting and badger digging. In today's climate of conservation, and the growth of public opinion against "blood sports", the participation of William Palmer in both activities, would not be acceptable to many people.

William Palmer followed the pack of otter hounds that used to hunt the Kent, in search of the shy, elusive creature that has now become so scarce in the Cumbrian lakes and river systems. "On our stream the angler and the otter are intimate acquaintances - I cannot say friends, though some of us are not inveterate enemies." As with cock fighting, maturity and an interest in animals for their own sakes changed his former boyhood attitudes. Neither did the young Palmer show sentiment in hunting and killing another of Lakeland's shy mammals, the badger. A prevalent attitude among farmers at that time, was that they believed the badgers worried sheep and lambs. They felt that this was justification enough for the persecution of this animal in fell country. William joined the groups of youths and men who searched the countryside to locate the active setts. This was a not too difficult task with an animal that makes its presence so obvious by the mounds of discarded earth that bank up in front of a hole. Once the sett had been located, men and lads embarked on hours of systematic digging into the series of tunnels that made up the badger's home. This was done in an attempt to bring the animal within reach of long handled iron tongs, by which the badger was hauled out of the opened sett. It mattered little where the animal was grasped by the biting tongs; a hold on its ear, face, leg or tail, all gave a the necessary purchase for the animal to be dragged out in the name of "sport." Only occasionally was the outnumbered animal able to get the better of its tormentors. In "Lake Country Rambles", William Palmer wrote about an incident concerning a gamekeeper, "at the conclusion of the chase, [he] thought he saw the animal's tail. Reaching forward, he took hold and essayed to draw. But he had made a terrible mistake. Seized by the paw, the badger whisked sharply round and reached the man's wrist. For full ten minutes it held him with his arm outstretched, and when finally it released its hold, the man's hand was hanging by a few shreds only."

William Palmer pursued badger digging with almost as much enthusiasm as he had for fox hunting, and he joined in many a dig into a badger sett that was situated in rocky terrain. This caused problems not only for the animals that were being hunted , but also for the terriers that were sent in to the sett to flush the badgers out. On occasions, the small , wiry dogs became trapped in the confined space of an underground tunnel, and were in danger from being torn by a badger's raking claws, and powerful jaws. There was also the very real possibility of the dogs being buried alive by rock falls that were dislodged by the frantic digging from above. The mens' excitement reached fever pitch as they encouraged each other in their efforts to dig and lever away rocks.

At the first opportunity, the dogs were recovered, and the tongs thrust forward to drag the badger out. Often the creature was not killed immediately, but the bloodthirsty activities continued by badger baiting. It was a common practise to physically disable the creature in some way before it was put in the confines of a pit , where it was attacked by a number of dogs. Wagers were brisk as men backed their dogs; the only certainty being that some of the animals died in the course of the savagery that provided the mens' "entertainment."

Today, William Palmer would stand condemned for joining in such activities; but his involvement has to be set against the way of life of a hundred years ago. It is sad to reflect, that even though modern legislation affords protection to badgers and their homes, these quiet creatures are still illegally tormented in the name of "sport."

Most of the pursuits in which William Palmer showed an active interest in his younger days, could be termed field sports, where his opponent was a wild creature. There were occasions however when he did match himself against the ability of other men on the arena or track.

In common with most of the village people,William Palmer was a regular visitor to one of the great attractions in the Lake District during the summer months, the Grasmere Sports. This was, and still is, one of the major athletics event of the Lake District calendar When William attended the sports, as a youngster, the programme included such event as wrestling bouts at all weights, fell and foot racing, long leaping, the high leap, and a one time, the pole leap. The Sports are traditionally held on the last Thursday in August

William Palmer first visited the Sports when, as a lad of ten in 1887, he walked to Grasmere from his Bowston home. To a ten year old, the all weight wrestlers must have seemed like giants. Clad in their traditional decorated white singlets, displaying their long johns that were over-topped with embroidered velvet trunks, they created an impressive line up before competing in their events. To the mind of a young boy, the heavyweight wrestlers were synonymous with the upholding of the law. Many wrestlers from the Kendal area discovered that success in the wrestling wring was a satisfactory and an acceptable means of entry to the police force. "In my childhood, fireside yarns often included the history of this and that hefty village wrestler who went into the police force and by skill and determination became the terror of wrong doers."

For William Palmer, that first visit to Grasmere was such a highlight that it became an event to look forward to every year. From that first visit, he continued to attend the Sports whenever he could, either in his role as a reporter for the local press, or merely as a spectator who was there for his own enjoyment.

Special "academies" that are still to be found throughout Cumbria are the training schools for those who follow the Cumberland and Westmorland style of wrestling. Competitions at different weights are still held at sports and valley shows even today, where "World Championships" are included in the programme. While most of the competitors taking part in modern competitions are from the north of England, there has been a gradual introduction of a truly international element as competitors from Iceland and Scandinavian countries now take part.

As in William Palmer's time, the rules are relatively simple. A hold must be maintained throughout the bout, and only broken by a fall when a part of a wrestler's body, other than the feet, touch the ground. This form of wrestling has a ritualistic style all its own, locked together, arms clasped tightly in an under and over style, the competitors sway, their heads nestled together, almost in a trance. Slow sideways steps shuffle in unison, a parody of dance; then, an upward thrust, shift of foot, a lightning move, and one wrestler is dropped to the ground.

William Palmer's early recollections of the successful wrestlers at Grasmere encouraged him to try the sport for himself. His first venture into wrestling came after the end of "a crippling thirty mile walk" to a sports meeting held in a distant valley. He entered his name as a competitor in the "all weights" class, a somewhat optimistic choice for at that time he turned the scales at eight and a half stone, and stood only to a height of five feet, five and three quarter inches.

He was lucky when the draw was made for the first round as an odd number of competitors had entered the competition. William was awarded a bye. His good fortune held for the second round, and he won through to the third round without the formality of making a hold. His opponent in that round was a wrestler of much greater body weight than his own, but of incomparable sobriety. The vast quantities of ale that his opponent had previously consumed rendered him somewhat unsteady on his feet. No great effort was needed to wrestle his opponent to the ground, for the drunken wrestler was obliging enough to slip, as they were "tekkin hod", and thus gave William the first fall. William was eventually awarded the bout and so was through to the final. There, his luck ran out,

Grasmere Sports Field about 1900.

Wrestling at Grasmere.

for he was up against a wrestler of greater weight and skill, who immediately laid William "flat on the sod" for two consecutive falls. There followed a thirty mile walk back home, the weariness of the journey however was somewhat alleviated by the fact that he had the consolation of second prize money in his pocket.

Spurred on by the acclaim given to the winners of the fell races at sports meetings, and reluctant to enter further bouts of wrestling, William Palmer had a fancy for taking part in these events. As a young man, he was quite slight of stature, and was of a wiry build rather than heavily muscled. In terms of today's athletics events, he would have been regarded as more suitable for the endurance events, rather than those of an explosive nature. As a fell walker of proven stamina and speed, he fancied trying to become a fell runner.

At nearly all the Lakeland valley shows and gatherings, "guide races" are included as part of the programme of events. These provide the opportunity for fell runners, of all ages, to compete in their respective classes. Most of the courses have the objective of reaching the summit of a fell, about 1500 feet or more, before turning for the descent and a triumphant return to the arena.

Success at this sport demands a combination of physical ability and courage. All the designated courses need the competitor to have ability at running, scrambling, and hands on knees walking over steep gradients to get to the top. This is followed by a hurtling descent over rough rock, scree, bracken and heather. Most of the races are won in times under the half hour. The "guides race" at Grasmere was, and still is, regarded as the premiere event in the fell running calendar. As the winner re-enters the arena, stirring brass band music acclaims the "conquering hero" as the runner heads for the finishing tape. In Palmer's day, winning a guides race at Grasmere gave an unknown village lad the status of a celebrity.

In an attempt to achieve this honour, William Palmer went for a few trial runs, in which he hoped to transfer his fell walking ability to a speedier ascent. However, the different techniques involved between fell running, and fell walking did not suit him. Even though he offered himself for training at one of the local fell running "schools", he did not receive much encouragement to continue with the sport. After demonstrating what he could do, he was advised to give it up. The trainer told him "a fell runner needs more than determination."

It was not until many years later, in the summer of 1947, that William Palmer was able to bask in the reflected glory of his great nephew winning the Youth's guide race at Grasmere. This gave him almost as much pleasure as if he had won the race himself. At long last, the name of Palmer was recorded as the winner of a guides race at Grasmere.

What did it matter if the name was Brian instead of William?

William Palmer congratulates Brian Palmer
on his winning run.

Chapter 8 - THE WRITER

All that has been written in the preceding chapters is relevant to William Palmer's writing career. Without all the knowledge and experiences that his many interests brought, it is doubtful whether he would have been such a prolific writer.

His interest in writing, allied to his natural ability with words, emerged while he was still at school. One of his nephews, Jack Palmer, recalled that "Uncle Will was allus writing when he was a lad, and allus won the composition prize at school." William Palmer was blessed with a good memory, and that, combined with an enquiring mind, enabled him to make the most of any opportunity that arose for recording items of interest. Throughout his life he kept diaries in which were recorded all manner of happenings and events. From these and a collection of other abbreviated notes, he was able, when the opportunity occurred, to enlarge or develop an idea into an article, that would be accepted by an editor of one of the many journals or newspapers to which he was a regular contributor

Although William Palmer wrote many books, it was his work as a freelance journalist that brought in his regular income. "He wasn't a wealthy man, and he had to earn a living somehow," explained a relative, William Walker, who at one time was the Mayor of Kendal. The state of William Palmer's finances was confirmed by John Chesterton of Ingleton, who was a friend of the Palmer family for almost thirty years. He had the feeling that during all that time, William Palmer was living on what he termed to be "the financial edge."

It is almost certain that Palmer's writing career began in the printing sheds of the Kendal based newspaper, "The Westmorland Gazette", although sadly no documentation is available to confirm that fact. He later transferred to the Northern Newspaper Syndicate, which at the time was a national and international press agency based a little further down Kendal's main street at 69 Highgate. His work at The Syndicate, with its small staff of three, gave him the opportunity to build up his own contacts in local affairs, business and the farming community, that were essential for him to obtain and supply interesting and informative material to a number of local and national newspapers, journals and magazines on a freelance, but regular basis. Some of these were quite prestigious publications and he indicated the fact that he was one of their regular contributors, by including their names on his personal note paper which was headed ; "Contributor to Spectator, Manchester Guardian, Contemporary, Field, Fortnightly, etc." While some of these titles have now disappeared with the passage of time, others still continue as household names today.

Occasionally, he wrote articles under the pen-name of "Dalesman", a title to which he could lay an authentic claim, for his childhood and youth were spent growing up close to the valley of Kentmere, while in later life, he was never happier than when exploring the valleys and high ground of mountainous country.

At the turn of the twentieth century, there were quite a number of independent local newspapers in Westmorland and eastern Cumberland to which he could contribute articles and features, but forty years later in 1944, he found that things had changed. After a visit to his native area, he wrote, "Forty years ago there were two weeklies at Ambleside as well as this Windermere issue, but nowadays the whole Lake District gets its news from Kendal [one paper where there were three] and Penrith [two instead of three]."

He was always on the lookout for material that would make an interesting item or feature. Like a squirrel storing nuts for the lean times ahead, anything that William Palmer wrote that was not of immediate use, was stored away for future reference. He kept the carefully typed pages in a home made filing system, which amounted to nothing more than a collection of boxes and chests, in which he stored all his reference material. He was not averse to supplementing his own pieces of work with cuttings that he had taken from newspapers or magazines, credited to other writers. If their subject matter was of interest to William Palmer, then it was likely to be investigated and represented in his own style. It was from these collective resources that he compiled many of his books and he admitted the fact in an introduction to one of his books, "I therefore ask the reader to accept this book as a compilation of pages, notes and sketches written over many years, and now revised and put into a permanent book."

This plea is frequently repeated but rephrased, in a number of his books and some of his writings appear in a slightly different guise in more than one publication, which prompts the reader to question, "Have I not read this before, somewhere?"

During the 77 years of his lifetime, W.T.Palmer wrote over thirty non- fiction books. The subject matter, in most of them, was of an exploratory nature and related to many different parts of the British Isles. The books were not produced at a steady rate, for the first was published in 1902, when he was aged 25, and the last was in 1952, just two years before he died. Many of his books were sufficiently popular and sold so well that they went into a number of revised editions, while the fifth reprint of his Penguin Guide to the Lake District came out in 1954, the year of his death.

There were however a few lean spells during this fifty year period, some of which lasted upwards of five years; during those times no book was published or revised. At other times three, or even more new books, or revised and updated versions of old ones, were brought out in a single year. Though their appearance in the bookshops was somewhat erratic, their production, if averaged throughout the period of William Palmer's lifetime as a writer, resulted in a book written and published every two years.

The contents of his books covered the wide range of subject matter in which the author was interested and familiar. Walking, climbing and his involvement with the countryside and its people formed much of the material. He reproduced anecdotes and recalled stories of customs and events, that when recorded, preserve for ever important information about local culture. How else can a native of an area learn about one's heritage if tales die out on old folks' lips?

He used a flowing, conversational style in his writing, which still make his books eminently readable. The subjects about which he wrote included history, geography, geology, botany, ornithology, to name just a few; and all were relevant to the Lake District, Scotland, Yorkshire, Derbyshire, Lancashire, Wales and Surrey. The contents of his books are still meaningful today although the passing of time is reflected in county boundary changes, technological developments, and changing attitudes and culture of the British people, at which Palmer would have wondered. His books are informative without being condescending, and they contain enough stimulating material to give a reader a base line from which to start his own area of exploration.

In the 1936 revision of possibly his most well known book, "The English Lakes", William Palmer listed his interests as "antiquities, fauna, flora, sports, geology, entomology, and the like", yet admitted that he was not sufficiently knowledgeable about any of the subjects "which might render my work of profit."

While there is evidence of meticulous research in his writing, there is no great scholarship exhibited in the contents of most of his books, or demonstration of deep philosophical thought. His books are very much of a factual nature, based on his comprehensive research, observation, and documentation about the lives and interests of people; as well as the background history and folklore of areas in which he was interested. Yet he displayed a command of language and imagery that brought the atmosphere and mood of landscape, in particular, right to the heart of his readers.

"It is a pleasing venture however, to struggle up Rossett Ghyll in thick cloud-mist, and at the top of Esk Hause to find clear air around, with the first flush of dawn rising, and being reflected in amber tints along a great sea of moving vapour. Then the chill of the breeze goes unheeded: columns and towers, and palaces of mist are being tossed up at every touch of the breeze into the golden light, and at last the level sunbeams play along the moving white tide, striking undescribable glories of rose and pink, and red and gold and silver, the while, steadfast, immovable, grey and brown mountains stand around."

For those who have had the good fortune to experience such a scene, W.T.Palmer's words exactly portray the magic of those moments. He encapsulates in a few words, the contrast between the elusive, ethereal qualities of light and mist, with the eternity and solidity of rock. He paints the scene with words of colour, he takes the reader along with him on his journeys.

Many of the events about which he wrote are so believable, because they could have happened to any of his readers, given that they share his enthusiasm, ability and capacity for exploration. There was nothing that he recorded that was beyond the grasp of understanding of an ordinary person, and this enables even a modern day reader to identify with the experiences of W T Palmer, and therein lies his credibility.

The writer Graham Sutton described his work in the following way, "His trump card was integrity; the sort that grew from belonging essentially to the life about which he wrote. Off comers tend to be no more than word painters of scenery - the still life men, in monotone or purple patch as their temperament moves them; if they want life in action they must research with diligence and report it at second-hand, having no other method

Dun Bull Inn, Mardale.

Loading hay - Loweswater 1913.

of arriving at the local truth. In this sense, Palmer did not need to arrive, because he was already there."

Most of his books were written in conventional chapter form, each of which is bursting with anecdotes, descriptions or accounts of events. He delved deeply into his own personal experiences, as well as recording incidents in which others were involved; yet all were relative to the central chapter theme. He wrote to this formula which evidently appealed to a wide readership, for many of his books went into a number of editions and revisions during the period of his lifetime. Yet now, out of print, and out of fashion, his books are only to be found of the shelves of second hand book dealers. It is sobering to reflect that the name of William Palmer, whose work was produced by still internationally famous publishing houses, is as unfamiliar to their present day editors, as to the man in the street.

But for those prepared to seek out the Palmer books, a treasure house of local and interesting information is there to be enjoyed. Even today, almost a hundred years after the first book was written, one can become sufficiently absorbed in a Palmer book to read it through from cover to cover; or alternatively by dipping at random into their pages, extract the little gems that emerge, bequeathed by different generations.

The books can be picked up again and again, to give reminders and insights into country people and their way of life, for the wealth of information contained between the covers, does not yield lightly to an initial, or cursory inspection. They make ideal reference books as they record in great detail, aspects of everyday life that existed, between fifty and a hundred years ago, in mainly rural communities. He wrote about our past, about our country, in the way it was.

W.T. Palmer gathered enough material to be able to compile three major series of books, in addition to writing individual, and other unrelated volumes, that gave specific advise on camping, hill walking, or guidance to visitors to an area. His "Verge" series records his explorations in great detail as he travelled through many different areas of Northern England, Wales and Scotland. In this series of books, the reader is taken on a sequential journey through the area that was to be explored in detail. History and legend; wildlife and folklore were woven into the tapestry of his writing, as the reader follows him from place to place in a logical progression of travel.

The "Wanderings" series was compiled quite late in his literary career, and all five books were published during the last ten years of his life. Their style, while still maintaining great topographical detail, is more personal and William Palmer discloses greater information about himself and his family in these five books, than in all the others.

The "Odd Corners" series is just that; a collection of snapshot passages based on a series of visits to villages and valleys, or of gentle explorations that he conducted on higher ground. When William Palmer wrote these books, in pre - car ownership days, all the locations to which he referred, were easily reached by public transport. Although this may be no longer available, the content of his books is still relevant, for most of the locations can be reached by car. These books are ideally suited to the would be adventurer, who likes the reassurance of the exploration having already been achieved.

During the years 1902 - 1952, when Palmer's books were published, the public transport network, on which he depended to get about the country, went through tremendous stages of development. It was extensive enough, and sufficiently reliable to enable him to roam far and wide, and while his "Odd Corner" series did not fit into a guide book format, he presented the information in such a way, that if so desired, his readers could follow in his footsteps.

Throughout his lifetime, he never had the inclination to learn to drive a car, which would have made it easier to travel about the country. Instead, he made great use of the fact that public transport was regular, reliable and adhered quite strictly to the set timetables. This gave him the reassurance that buses and trains were available, not only at the start of a day's exploration, but were there at the end of the day, when tired limbs ached for the journey home.

Although relatively few people were to be found enjoying walking, climbing, cycling and camping, for their leisure pursuits, in the early part of the twentieth century, the end of the first world war brought about a great upsurge of interest in all things relative to the outdoors. John Chesterton, who was a friend of the Palmer family, feels that William Palmer was very much a man ahead of his time by being involved in and writing about these activities. No other writer of his generation could match his prolific output of material that related to the outdoor life, which ultimately began to be enjoyed by a large percentage of the British public.

William Palmer's great love of walking in the familiar countryside of his native county, provided him with much subject matter in the early days of his writing career. He was a countryman with an eye for country details. He made note of the birds and flowers he found along the way; he described the routes that he had followed on his way to a mountain top, or the countryside through which he walked or cycled. He took time to talk to local people, listening to stories told by farmer or shepherd when he stopped for a bite to eat at village inn. He listened; he remembered and he wrote down what he heard.

His first book was published in 1902, at a time when it was just becoming possible for people, of limited income and leisure time to visit the Lake District. Until then, the nineteenth century visitors had been mainly an influx of scholars on reading or study vacations, and their numbers were increased by people of private means, who came to enjoy the peace and beauty of the Lakes for periods of weeks at a time.

Before the development of the country's railway network, travel to the Lake District was by horse drawn coach. Passengers had to make their journey in a series of stages that linked the south of England with the remote north. In 1829, 12 different coaches arrived in, and departed from Kendal on a daily basis. They brought early visitors to the district, encouraged by accounts of the romantic Lakes, as described in guide books such as that written by William Wordsworth. Later years such as Jonathan Otley and Harriet Martineau were also to extol the beauty and attraction of the Lake District.

In 1847 the railway began to bring visitors to Windermere, the trickle developed into a steady flow with the onset of motor transport, when their journeys by road and rail gave an introduction to the Lake District. This was the background that heralded the onset of tourism and commercial exploitation of the Lake District. For most of those early visitors to the Lake District, their presence was brief. A day trip, a weekend visit or at the most, the luxury of a week in "digs", was all the time the holiday makers had to spend away from the industrial scene. But visits, however brief, sparked an eagerness and enthusiasm to know more of the area they left behind. There was a need to know more about the places they visited rather than just express admiration for the beauty of the landscape, and William Palmer was the man to tell them.

One reason why his books and articles appealed to the visitors, was that they were written by an ordinary man, with an understanding of ordinary people; written in language that was easily appreciated. There was none of the mysticism or symbolism favoured by some 19th century writers, where a reader is left to interpret the writer's ideas and thoughts. The work of Father Thomas West, Jonathan Otley , and Harriet Martineau had greater appeal to the scholar or person of substance. William Palmer's writing was attractive to the masses, in much the same way as the distinctive style of Alfred Wainwright had such popular appeal to a later generation. William Palmer's writing was straightforward and to the point. If colour and imagery were needed to enhance his account of a scene, he could provide it, while on other occasions, events were left to speak for themselves. His style of writing varies from the crisp, informative reporting style of a journalist, to the vivid descriptions of one painting a picture with words. His writing was simplistic in style, yet remains infinitely readable. He made no pretence to scholarship, but through his writing, he satisfied his need to pass on to those that he regarded as not as fortunate as himself, the delights, adventures and experiences that enriched his life. William Palmer was aware of the fact that many of his readers had neither the opportunity, or physical ability to encounter at first hand, some of the situations he experienced; yet why should they be denied a share? Through his writing, he took them along with him.

"In winter the mists are horrible. I don't suppose many of my readers have ever crossed the desolate snow covered uplands. It is dreary enough work when the pallid sun glints along the even surface, lighting up the air with an unwonted shimmer, and the great crags loom out on the fell sides." The supposition having been stated that his readers are unfamiliar with what he has experienced, he then introduces the pronoun "We", and takes us along with him to experience the rest of the crossing of Scarf Gap in a snow storm. We share the journey with him; we see through his eyes the magic of Lakeland in winter.

The reviews of William Palmer's first book were favourable. Of "Lake Country Rambles", the critics wrote, "The book has the great living quality of faithfulness. The author renders simply but vividly just what he has seen, and, as his experiences have often been adventurous, this earnest accuracy results in a narrative power which constantly hold the reader. There is quite an Homeric flavour about some of the pieces." [Speaker]

Climbing on Kern Knott, Great Gable.

"Mr Palmer possesses in a very eminent degree that true art which consists in dressing nature to advantage, and it is joined to a clear, natural and picturesque style." [Glasgow Herald]

The Morning Leader described it as "Mr Palmer's charmingly written volume. If one would know what magic this wonderful district holds he must go to Mr Palmer for further guidance."

"Mr Palmer must not be alarmed if his reviewers envy him. Those "Lake County Rambles" of his have a keen delight that a saint might covet" [Daily News], and so the reviews went on.

His next book "In Lakeland Dells and Fells" was published the following year, in 1903. Much of the content of this book was directly related to his own personal experiences and involvement in Lakeland sports, and exploration. He also reproduced tales that were passed on to him by country folk, or stories that he had discovered while reading older books or documents. For the spoken tale, he often included a credit as to its source, "I wish it to be clearly understood that I am reproducing without ornament or argument, the tale of a mountain catastrophe as told by a rheumy little man of sixty-five, the holder of a well known sheep farm among the fells." The same degree of credit was not always attributed to written material. Some of the stories, or information that he discovered in older books are either written in their entirety , exactly as he found them, or their content reproduced in his own words.

Yet he was not infallible in his recording of events. In the Kitbag Edition of his "English Lakes" , when writing about early rock climbing in the Lake District, he attributes Walter Parry Hasket Smith's first ascent of the Napes Needle as occurring in 1884, when in fact, Smith's own recording of the event in the Wasdale Head Visitor's Book was for June of 1886.

William Palmer wrote, "In 1884 with John Wilson Robinson, he located the spike of rock on the great cliff, [Napes and Needle ridges], and a few days later it was conquered - a wonderful single handed achievement." Smith did indeed locate the Needle in the company of John Robinson in 1884; which was a rediscovery of finding the rock two years earlier, but in 1884, no attempt was made to climb the Needle. So apprehensive was Robinson at his first sighting that he queried whether a Swiss guide would be needed to assist them in the ascent.

Is it possible that Palmer made a simple mistake in the use of a "few days", instead of a few years?

This is highly doubtful, for in an earlier paragraph in the same chapter of the same book, he again uses the date of 1884 for Haskett Smith's great achievement. "In 1884 it was possible to make a first ascent of the crag on the Ennerdale side, a descent of Needle ridge, and a conquest of the Needle itself, all first ascents, in one afternoon."

Lehmann J Oppenheimer, a contemporary of William Palmer, confirms in his book, "Heart of Lakeland" that in 1884, Robinson and Haskett Smith did make first ascents of the Ennerdale crag and the exploration of the Needle ridge, but when they reached the

Needle, which was new ground to them both, they declined to climb to the summit at that stage. Instead, they contented themselves with "threading the Needle", by climbing behind the pinnacle, and the parent rock of Great Gable.

What makes it so amazing for William Palmer to make this mistake in 1930 , is that while he may not have had access to the Wasdale Head Visitors Book, which contains a cryptic record, by Smith himself, of his ascent of the Needle, Palmer did have knowledge of Haskett Smith's own account of the climb in an article which was written for the Fell and Rock Journal of 1914.

The editor at that time was W.T. Palmer.

An odd mistake with dates can easily occur, but for a mistake to be made over such a momentous event in the annals of mountaineering as the first ascent of Napes Needle, is remarkable. But putting Haskett Smith's achievement into the perspective of William Palmer's viewpoint of over sixty years ago, it may not have assumed the importance that it does today. When history is being made, the makers are often unaware of its making. Hindsight and posterity nominate which events should attain importance and acclaim in the greater scale of time.

Perhaps I judge William Palmer too harshly on this matter. After all, he made a solo ascent of the Needle only a few years after that first climb. When set against Palmer's own success, maybe Haskett Smith's ascent did not carry the same accolade of achievement with which it is acclaimed today.

William Palmer became editor of the Journal of the Fell and Rock Climbing Club in unusual circumstances. Early in 1910, E H P Scantlebury who was then its editor, proposed that the name of the Journal should be changed to the Lakeland Journal. The proposal was put to the members of the club, but was defeated in the vote. Scantlebury resigned, and William Palmer was elected in his place. He held the position from 1910 until 1918 during which his period of office was fraught with difficulties. Delays in publication of the Journal were caused not only through the inevitable difficulties imposed by the restrictions of the war years, but also problems with his own personal health.

He quickly learned that the honorary position of Editor was no sinecure; it was a year round job. Articles that had to be considered for inclusion in the Journal were sent to him from all over the world, and the Palmer kitchen table was recorded as being "under a mound of paper and correspondence for the whole year." During this period of editorship, the losses of men and the horrors inflicted on the battlefields of France, affected Palmer emotionally. In the 1915 issue of the Journal, which he entitled the WAR ISSUE, he published a number of letters sent to him from Club members who were on active service. Memorials to others appeared in the obituary section of the Journal, where tributes were paid to those Club members who were killed in action. William Palmer ensured that, wherever possible, any information of members engaged on Active Service, was passed on to others by way of the Journal, a copy of which was sent to every known address. William Palmer felt that by conveying the thanks and support of fellow

The Forgotten Man of Lakeland

William Palmer

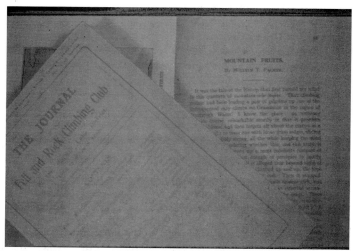

Editor of the Fell & Rock Climbing Club Journal
1910-18.

club members to those fighting in the war, it maintained some degree of normality in an otherwise abnormal world.

While he was the editor of the Fell and Rock Club Journal, he had access to a great deal of interesting material that was submitted for potential publication. Lack of space or unsuitability of content matter meant that not all of it could be used, but they were not forgotten by William Palmer. Some pieces, even those already published in the Journal reappeared years later, in Palmer's own books. In a number of cases this was done without even crediting the original source. A case in point was an article entitled "The Doctrine of Descent" by S.W. Herford which appeared in the Fell and Rock Journal of 1914. Much of this article is reproduced under the same title as part of a chapter of Palmer's own "Complete Hill Walker" which was published some ten years later in 1924; without any acknowledgement made to the original writer. Other pieces also have that "deja vu" feel about them, like his accounts of Auld Will Ritson at Wasdale Head, that bears a striking resemblance to a piece written by George Seatree, who knew Auld Will personally. Palmer openly admits to using material from sources other than his own, and in some of his books he does apologise for lack of acknowledgement where the source is unknown. In the introduction to More Odd Corners in English Lakeland he writes, "I have not been able to trace the origin of every chapter in this book. Some indeed are written from information taken from many magazines and review articles, and to the editors of these, should my offence be discovered, I offer my apologies. The liberty is taken so that the results of many years of research and writing may be available to present day ramblers and readers of Lakeland books."

In instances where he used material that had been submitted during his editorship of the Fell and Rock Journal, he would have been aware of the origin of pieces that were sent to him. Even after his resignation as editor, he still received a copy of the Journal for each year of it's publication, and therefore he should have been familiar with its contents. Had William Palmer been writing today, under the strict code of practise of the copyright protection laws, he might have been taken to task about the sources of some of his material.

Yet the point he makes about preserving researched material is a valid one. As each generation emerges to the curiosity and interest in their own background, it is usually to the newly presented material that many of them turn for their information. Where credit is known to be due however, it should always be given.

In 1905, William Palmer produced the text for what is probably, his most well known book, "The English Lakes." The publishers, A & C Black commissioned the Grasmere artist, Alfred Heaton Cooper to provide the illustrations for the book. Like W.T. Palmer, Heaton Cooper was an enthusiastic mountaineer who loved the Lake District. The days that he spent among the mountains provided him with his subject matter, and Heaton Cooper's illustrations and William Palmer's words blended perfectly in what has become a Lakeland classic.

Yet, according to the artist's son, William Heaton Cooper, it is unlikely that the two men ever actually corroborated on the publication. Both were commissioned individually by the publishers, for the production of the book, yet although they worked a individuals, the blend of Palmer and Cooper was a perfect. The book was so well received, and in such demand, that it was constantly revised and updated through many editions. Although the text changed slightly with each new edition, the original illustrations remained unaltered.

Some of William Palmer's later books were illustrated by photographs taken by another Kendal based man, Joseph Hardman. Like William Palmer, Joseph Hardman was a great explorer of the Lake District, but unlike the writer, Hardman rarely strayed far from the road side. His explorations were done from the comfort of a hired car. William Palmer's granddaughter, Christine Buchanan, explained that of all the photographers who illustrated her grandfather's books, he liked working with Joseph Hardman most of all.

Hardman was a photographer who had the great skill and ability to capture a quality of light in his photographs. He had the mark of all great photographers to stamp his own style on the images that he produced, and which are easily recognised as Hardman's work. This may have been the attraction to William Palmer, for while he tried to capture and convey in words, the moods effected by the changing light on the landscape, Hardman's photographs caught that same atmosphere with the pressing of a camera shutter.

William Palmer was very aware of the changes that the galloping progress of the twentieth century was bringing to areas about which he had written, and he was constantly updating his material. Even when working on a new manuscript, he was always ready to insert the latest information that came to his attention. This was another indication of the journalist's awareness of the movement of events. His granddaughter, Christine, who lived with him for a while, could recall that he always seemed to be busy rewriting or updating one of his books, while yet working on another. "I lived with Granda at the age of eight, and could not then have known precisely what he was doing. My recollection is the constant clatter of a typewriter, and the injunction upon myself not to be too noisy. My Aunt Jean explained that he was working on the revision of one of his books right up until a few weeks before he died," she recalled.

Much of the content of his books was inevitably based on his own personal experiences, and it seemed as though incidents about which he wrote, would jolt a spark of memory into life, enabling him to recall some other half forgotten incident. It was almost as though his mind was a labyrinth of adventurous ways, of fascinating corners, of curious depths, which yielded their secrets through his typewriter keys. William Palmer was never far away from his typewriter if he could help it. If he was away from home or his office for any length of time, and could take the machine with him, then he did so. Even when he was involved in a walking expedition, then the typewriter was taken along. He ignored the extra burden of its 10lb weight for the rewards that it brought him.

In his later years he travelled about the country by car, which was driven by his daughter Jean. She always had to make sure that the typewriter was packed in an attache case, along with items of stationery that were needed for her father's work. The case was placed, easily to hand, in the back of the car.

Although many of his books were compiled from his already collected material, others were as the result of planned expeditions. He would go away from home for a few weeks at a time, during which he followed his carefully constructed itinerary in its logical sequence. He always allowed extra time so that he could capitalise on any interesting diversion that came along the way, for he was nothing, if not an opportunist. To William Palmer, the unexpected was likely to be expected. He seemed as though he could anticipate events; he was a man to whom things happened. The number of occasions that he was involved in unusual incidents or events was more than mere result of chance. William Palmer had the knack of being in the right place at the right time, for he had a fair idea of where the right places were, and was prepared to seek them out.

He was shrewd enough to know that a wandering visitor who arrived at a lonely farm or cottage, was an unusual event in itself to those who lived there, and the result of his visit was likely to yield some item of interest. To farm folk starved of company, a visiting stranger was a welcome diversion from the monotony of farm life. It provided the opportunity for new conversation, to exchange ideas and philosophies, and to pass on stories that had been handed down through generations of the family. All this was potential information that could be used by William Palmer to good editorial effect.

During these lengthy, planned expeditions, William Palmer allocated two or three days to explore each area. He used this time to gather information about events of importance, customs and folklore, and to record the wildlife that was to be found in the district. He made notes on the spot; recorded details of his finds, and when he gauged that there was sufficient material to make a halt worthwhile, he stopped in his tracks. This halt to the expedition enabled him to devote a full day to writing his articles, or chapters of his books, when he transformed his rough notes into a finished piece of work. On occasions, especially if he was working to a publisher's commission, the articles were posted en route, to the waiting editor.

Working in this way, classified William Palmer as one of the early professional explorers, albeit in his own country. His expeditions were planned in such a way that they were virtually self financing. He was very conscious of having to meet editorial deadlines while on his travels, and on occasions, when time was pressing, he almost found himself at a loss for words. "Much of it was hard graft for me; editorial wishes had to be strictly observed, and always there was the effort to keep a keen outlook for the unusual."

While away on these lengthy trips, which were sometimes of a few weeks duration, it was just as important for him to be contacted, for both family and business reasons. Keeping in contact with the wandering William Palmer was sometimes an arduous business, as much of his journeying was spent well away from the beaten track. Without the benefit of today's mobile phones, he had to rely on postal messages left for him at designated places, such as post offices or hotels along his route. Usually he was very conscientious about responding to them, especially if they were of a family matter, but occasionally he chose to ignore any messages from his publisher, going so far as to give specific instructions to his family. "Do not send on any messages. Skeffington's can wait for the index."

To relieve himself of the burden of carrying completed scripts, they were sent either directly to his publisher, or in a letter form to his wife who would keep it until required at some later date. These letters he sent were often lengthy affairs, as was one that was sent to his wife from France in 1923. This was in the form of a daily journal, which only emerged during the course of researching the Palmer material for this book. It was discovered by his granddaughter, tucked away in a box of personal family items. From its nature, it is almost certain that it was intended to form the basis of a book, for it contains a comprehensive account of his stay in Paris. The diary indicated that William, accompanied by his elder daughter Annie, had gone to Paris on a guided tour. The tour appears to have been a combination of a mind broadening experience for Annie before she commenced her university studies, and an information gathering expedition for William. The diary was written in the form of a letter to his wife, who at that time was spending a holiday in the south of England. The little book, with its tightly compact writing, contains a comprehensive record of almost every movement that William and Annie made from their stomach heaving journey across the channel, to their relationships with their fellow travellers. It can be read in full in a later chapter.

Through the pages of the diary, William Palmer reveals different aspects of his writing and personality in this very personal document. The shutters of commercial writing are drawn aside to reveal a man of great humour, and self mockery. His diary indicates that he was not greatly impressed by the Paris streets and its traffic.

"A million English folk have been to Paris - ah yes, and they do again that the glory of that city is in its streets - wide boulevards radiating like spokes of a wheel, and a roughly circle of roadway right round the city. Now comes one man with an eye and memory. There will be deuce to pay for this.

Paris as an ideal street city is a delusion, the actuality is not a dream, its a nightmare. I grant there are boulevards as stated, but two thirds of Paris traffic drives unregulated through streets full of twists and bad corners at a fiendish speed. I don't care to ask about casualties."

The fashions of Parisiennes that he observed on an evening stroll also proved to be something of a revelation to him. "Bare arms and necks and backs flashed everywhere. Three out of four women rouge their cheeks. Four out of three are carmine with lip salve. You get a great flash of it fifty yards away, and it looks worse every step you get nearer. Its a sorry show, but when Mary Jane rubs her top note with black lead round her eyes, and boot polish for her eye brows, the offence screams for shadow against the sun. I don't guess what happens about their hair: its lack lustre stuff like hay or straw at best. Parisian ankles ? My hat ! I have yet to see any worth noting. The present fashion decrees loose dress round the waist here, and no corsets. In a motor the women look fairly good but stout. When they wobble on two poor feet along the street, I've got to laugh."

The diary is a gem of observation and attention to detail. Although William Palmer indicated on more than one of its pages that part of his intention in visiting France was to write some commercial material, the production of that actual work escaped him. It's a great pity that no commercial advantage was taken of the visit, for had a manuscript that was composed in the light hearted vein of the diary, been submitted to his publisher, some fascinating "Odd Corners of Paris" may have been preserved for posterity.

William Palmer was a perceptive man, and it was unusual for him not to seize on this opportunity. Normally he was quick to see the possibility of turning any incident to profit. At home, in his native country, no time was wasted; a chance remark, or a casual glance was never ignored , for these were often all that was needed to set up a few hundred words of editorial copy. Even a train journey that William Palmer made through the Border region, and into Scotland was turned to good effect, for he recorded details of what he saw from the carriage window.

"Beyond this gorge, there is flat moor, and the Allan is a winding, slow, and today overcharged stream. The holms along the water had a large number of young horses so that this form of stock breeding is going ahead. The cattle which were Shorthorns in Cumberland, of the Ayrshire type in Dumfries, are now more and more of the black Angus sort, many with white lines and patches underneath."

Such a report may seem somewhat mundane, but the value of articles of this nature, which were written with attention to detail, is in the actual documentation of an ordinary scene. Accounts such as this form a useful record of the changing nature of the countryside, and its agricultural industry. The Shorthorns and Ayrshires of Palmer's day, have now given way to the continental imports of Charolais and Limousin, in the farming quest for greater productivity.

There is no doubt that William Palmer made a success of his writing career; for with his enquiring mind, and natural interest in people and the countryside, it seemed as though his supply of subject matter was inexhaustible.

It was his work as a journalist that decreed where William made his home. In the 1905 edition of Bulmer's History of Westmorland, William Thomas Palmer is listed as living at 5 Lake Road Terrace, Kendal. His occupation at that time was given as a compositor. He was then aged 28 and already had three books to his credit. In 1918, he moved to Liverpool, to work on a regular basis for the Liverpool Echo, the Liverpool Daily Post, and the Liverpool Daily Courier. The editor of the Courier was Lewis Chesterton, who shared Palmer's love of all outdoor activities. This was a move that brought dividends to the Palmer career, for Lewis Chesterton was a man of some standing in the city, and was able to introduce Palmer to other freelance outlets that brought work his way during the difficult years of the twenties and thirties.

The period from 1918 to 1930, was a lean time as far as William Palmer's book production was concerned. These years, following the end of the first world war saw the country enmeshed in a period of poverty and depression, as the country and the people struggled to rehabilitate after the difficulties of the war years. Money was scarce for both consumer and publisher, and the necessity for many families to live on a tight budget, ensured that priority was given to the supply of basic necessities only. Food, fuel and clothing were more important than reading material, the purchase of books could wait for easier days. Newspapers however, were still regarded as essential reading in those difficult times, and William concentrated on this aspect of his work, as well supplying articles to cycling and camping magazines that were beginning to appeal to readers who clamoured to escape from the drabness and poverty of their lives, to the outdoors.

This was a difficult period in the industrialised areas of Lancashire where both work and wages were in short supply. Tempers were short, and violence simmered below the surface of work hungry men. A nephew of William Palmer, William Walker recounted how on one occasion, an article written by William Palmer, that appeared in one of the Liverpool papers, helped to cool rising tempers during the general strike of 1926. William Palmer was aware of a potentially explosive situation that was developing between the strikers and the police. He spent all night at his typewriter, composing a conciliatory article that he hoped would help to calm things down. It did.

In his writing, the character of William Palmer comes through as being a quiet, somewhat reticent man. This was borne out by his nephew, Jack Palmer . "He didn't make a song and dance about anything; he didn't boast about anything. That wasn't his way at all." Another Kendal man, Kenneth Shepherd, who for many years was a photographer with the Westmorland Gazette, only knew of William Palmer in his later years. He also described him as a reticent man who did not advertise himself.

This may have been a side to the Palmer character that was familiar to his family and friends, but he presented a different face to the hardened business world. William Walker who stayed with the Palmer family on many occasions in Liverpool, saw him in a this light. He agreed that usually William Palmer was a quiet man, but that was not the side of his personality he presented when meeting with publishers or editors to whom he wanted to sell his work. "I've been with him when he wanted somebody to use what he had written, and he knew how to push himself then." This assertive side of the Palmer character was confirmed by John Chesterton, a family friend who remembered that William Palmer could be very forceful in the pursuit of his journalistic career. He also recalled that William Palmer was an ambitious man as far as his writing was concerned, and he was prepared to promote himself in order to get a commission. When he met someone for the first time, especially if there was a possibility that that person could be of use in furthering the Palmer literary cause, William made it clearly known that he was the man for the job. He set out to deliberately sell himself to a potential employer by pointing out what he had done, and more to the point, what he could do. He advertised his own ability, and if the need arose, was prepared to enhance his own image. He adopted what in today's terminology, would be described as a positive approach.

The years spent living in Liverpool gave William Palmer the further opportunity to explore other parts of northern and central England and North Wales. Naturally, these visits led to the production of additional books to his already extensive list. "Odd Corners in North Wales" was followed by "More Odd Corners", which like his Lakeland "Odd Corner" books, gave would-be explorers information on interesting sights to see and places to visit, without straying too far from the beaten track. Another book, "Wales Its History and Romance" was published in 1932, and this is probably the most scholarly of his books. It is a well researched and comprehensive account of the history and development of the country and its people.

The Forgotten Man of Lakeland

In 1934, William and Annie Palmer moved back into the Lake District to live at High Wray, near Ambleside, from where he was still able to carry on with his freelance writing for journals and papers. During these later years that he spent in his native county, he extended his explorations further into Cumberland, where he travelled the extent of the west coast, discovering items of interest and architectural gems tucked away from the popular tourist area of the Lake District. Even today, this is a somewhat neglected and undiscovered part of the county, as far as tourism is concerned. But in the pages of Palmer's book, "The Verge of Western Lakeland" which was published in 1941, he explained to the nation at large what the residents of West Cumberland already knew; there was just as much of interest in that region as to be found in the central lakes area. The only difference being that one may need to look a little harder to find it.

Living so near to Ambleside, with easy access to the Scottish Borders, enabled William Palmer to plan more expeditions north to collect material for his Scottish books. "The Verge of Scotland", was published in 1939, and he was also able to compile the preparatory work for the "Verge of the Scottish Highlands" which was published in 1947. The house to which William Palmer had moved from Ambleside to Kendal in 1939, was on Cliff Terrace, and was bought from Darwin Leighton, a fellow member of the Fell and Rock Club. By this time, if Palmer was not exactly a wealthy man, he must have been of comfortable means, for the substantial, grey stone house that overlooks the town, still has an air of quiet gentility about it. The house has maintained its connections with the Palmer family over succeeding generations, for William Walker has lived there since about 1946, when the Palmer family returned to live in London. He recalled that when William Palmer lived in the house, the walls of the sitting room were lined with books, many of them first editions that were sent to him by London publishers for review purposes.

"He was a terrible man for books", William Walker commented.

It was mainly as a result of the publication of his book, "The Verge of Scotland" that he was elected a Fellow of the Society of Antiquaries [Scotland]. The current secretary, June Rowan explained that while the majority of Fellows are eminent scholars in their specialised fields of history or archaeology, this was not a prerequisite of Fellowship in William Palmer's time. An interest in Scotland and all things Scottish was then the major requirement, and his book with its many historical and antiquarian references made him eligible for Fellowship. He needed a sponsor, but no record is available as to who that was, and once elected, the payment of an annual fee was the only requirement to append the letters FSA [Scot] to his name.

William Palmer still managed to combine his journalism with review work of newly published material. He was also sent manuscripts to read that had been submitted to London based publishing houses by hopeful authors. He continued to write his own books, but his nephew felt that he lost credibility with some of his later titles. "The books he wrote about the Lake District were genuine. He knew the Lake District like the back of his hand. There wasn't a bump or a hollow that he didn't know." He accepted that the Welsh, the Scottish, and other books about the north of England were also written as a result of first hand experience gained through exploration allied to research, but he felt the books about the south of England were what he classed as "potboilers." These, in his opinion contained too much information that was taken from other sources, and reproduced to the formula that William Palmer had already successfully established. As a result, his nephew felt that his reputation as a writer suffered, and this aided, "his fall into obscurity, but he doesn't deserve it."

William Palmer himself refutes the claim that his southern material was ghosted from other writers, in his introduction to "Wanderings in Surrey" which was published by Skeffington shortly after the end of the second world war. In the Preface, he writes, "To Wanderings in Surrey, the attention of five years has been devoted; it is written entirely from experience. Every town, river, common, ridge wood and road in the county has been visited, with most villages and hamlets. Much of the first draft of the MS was mislaid, (by myself); it had to be rewritten and the completion is delayed on that account."

The onset of the disease which was to claim William Palmer's life had already begun in the 1940's during one of his most prolific writing periods. Yet, during the last ten years of his life, twenty four new or revised editions of his works were published, and this was in spite of the mild symptoms of forgetfulness, and loss of memory that were surreptitiously beginning to appear.

Inevitably, the question is asked, "Why did William Palmer fall into obscurity?"

The books that he wrote are still eminently readable; the mountain landscape does not change a great deal, so his descriptions are still as relevant today as when he wrote his books, and the folklore that he recorded is part of our countryside heritage. But in Palmer's day, there was no commercial advertising on the scale as we know it today. If people are unaware of a product, they won't buy it, however good or interesting it may be. Even the Wainwright Guide Books that have sold over a million copies, received a tremendous boost after the author's exposure in a series of television programmes.

Advertising, and marketing, accompanied by a series of interviews on television or radio by an author, are all designed to enhance the sale of a book. Without this assistance, and the still relatively small numbers of visitors that came to the Lake District in the utility years following the end of the second world war, the demand for Palmer's books was no longer there. Many of the firms that published William Palmer's books also disappeared, as they were taken over by larger publishing houses.

But as the Palmer star declined with his death in 1954, that of another Kendal based man was about to rise. In the year following William Palmer's death, Alfred Wainwright published the first of his Pictorial Guides to the Lake District, which became an overwhelming success, and in some way may have obliterated the need for those written by William Palmer.

It is interesting to speculate how William Palmer's books would sell today, if they were given an updated presentation, backed by the power of a modern advertising campaign. The information they contain is still meaningful; it is only their faded backing boards and dusty pages of a former age, that consign them to the past.

Chapter 9 - A VISIT TO FRANCE
by
William T Palmer

In the summer of 1923, five years after the end of the first world war, William Palmer took his eldest daughter to Paris for a week's holiday. They travelled as members of a small organised party. Annie was nearly 18 years old at the time, and it is possible that this holiday was partly to broaden her experiences of life before she began her university studies.

The diary is reproduced in full, and gives a comprehensive description of the city through the eyes of a patriotic Englishman, who had evidently studied the history of France in general, and Paris in particular, before he embarked on the holiday.

Although the diary, after its completion, remained in his possession for almost 30 years, and it was obviously his attention to use the material in some way, there seems to be no evidence of its publication.

Over seventy years after his visit, with day travel between London and Paris available by rail, car, sea and plane easily available, the diary gives a fascinating insight into the impressions that Paris made on William Palmer.

Paris, France,

19:8:23

My dearest,

To follow on my Friday's screed. We reached Victoria in excellent time: met Dean & Dawson's man, and got first rate seats on the train. Annie had slept very well from about 9.45 to 7 am. She is very fit.

The run to Newhaven was pretty but not striking, except that the old Castle at Lewes stood out finely. Anne got a rough pencil sketch of its outline. At Newhaven the wind was blowing a strong breeze. We took the SS Versaille, a French boat, and on advice of our guide had lunch before she left harbour. It was as well. In the channel the boat began rolling at once, and kept up the exercise until we reached Dieppe, a good three hours. St. Eloises was not in it. The waves are short broken and lively. We had a big crowd on board, and casualties as to sickness were numerous. Annie and I got a view of the sea, and Annie says she gave her lunch up to the fishes. It cost us 25 francs including tips - or 6/3; a franc was 9 1/2d before the war; it is now 3d. I don't expect to find any difference in the cost of anything compared to England.

There was no error about the Versailles. She rolled until the water was nearly up to her rails, and despite all hanging on, the people and hand luggage slid down the deck. I had jammed my bags against a ventilator and Annie and I had got a nook in the winch, out of the way and safe, unless the whole ship went over. There is no error about Albion's white cliffs, about the fascinating colours of the Channel water, purple and dusk, bright green and blue. And the waves go all ends up. Brittania began to use her destiny; here she ruled the waves. But Mr Baldwin has forgotten the trick: he wants a brush first and then a comb!

My tummy was doubtful; I kept near the rail. What decided me against parting with my hors d'oeuvres was the fact that some feet to my left three exquisite ladies were jamming themselves against a stack of fish boxes from Aberdeen, exuding soft ice and, doubtless . I'm not surprised Dieppe folk import Aberdeen fish; they know what the Channel sort feed on. Finally, in mid ocean, our captain grew desperate. He had no coals and began to fire up with German marks, or the contents of his. A fragment, probably a millionth of a mark pasted me in the eye, and I didn't like it. Also we had a few gouts of spray, but nothing to matter.

Then for France - more chalk cliffs, and some pitching misery when the tide caught up outside the harbour. Then came the final catastrophe - to the umpteen lunches dumped to the fishes was added a gentleman's hat - not mine. With it went our crew's combined blessing that the Queen of the merrymaids might accept it as a crown, and no bon Brittanica, take a little more care in the design of jazz waves.

The whole of Dieppe and two visitors from Paris met us at the pier head. No doubt we had had a crossing, but as Columbus steered through the gap between the piers he found that all was well. The great continent of Europe was still anchored in the same place.

From Victoria I had been planning a desperate crime - not murder or pitching the cook overboard (which is the correct thing to do after you've mislaid your lunch), but SMUGGLING. Behold in this the confessions of a bad bold man, piracy is my worst offence. Another member of the party had three boxes of matches. The conductor warned him that Douanes would pinch one of them. I said no: I would be a smuggler. If any pinching was to be done, I, with the blood of Border cattle thieves, would do it.

Douanes didn't care. I brandished my box of safeties - he recited the alphabet backwards and scrawled 2 on the bags I wanted to open. I complained that I wanted to be chased, arrested and imprisoned in a distant foreign land, but he gave me a dissertation in Sanscrit, which Annie translated . "What else do you want?" I was disappointed. After the whole scene had turned out, he would not lift a finger and send a red - legged man to capture me.

French railways are understood, I hope, of French folk :they built them. Our train was half a mile down the quay, and wanted some finding. There was the usual comedy, but nothing to shout about. The street folk were wonderfully French, and their roofs has colour enough to beat jazz.

I'll skip the journey to Paris except to say that we had tea on the train. Annie tried chocolate, and we had as many spills as drinks. The permanent way is fiendish. We had French bread baked, butter without salt and pretty good jam.

Our entry into Paris was hectic. I don't remember such a scram in my life. They are rebuilding St. Lazare station.

Armistice day in Liverpool was a faint idea of a jostle and turmoil. The motor drivers and conductors shrieked abuse until the very taxis sweated. Why a thousand people in a day arn't killed, I don't know. And our conductor said that it wasn't bad. I'd prefer a crossing on the Hirondelle before I'd risk my life afoot in the melee. The scene is historic, and I don't wonder that climbers unsheathe their ice axes in Paris on their way to the Alps. I'd do it myself. Squeeze and push with two handbags for half a mile, in and out of barriers, is more than a joke. The crowd was almost solid.

In the yard, things were worse. We got a motor bus to the hotel, and inch by inch, the tangle of cars straightened out. In theory French drivers keep to the right: in practise they cut and drive any way, any time, and everywhere. Crossing the streets is a great adventure.

Our party is about a dozen in all; 2 from Stockport, 4 at least from Manchester, our 2 from Liverpool, and a nondescript lonesome who talks a sort of semi- Scots and lives in London. He is actually an Irishman.

The dinner at the hotel was very welcome: the food is always on the light side, but our table crockery is a bit heavy. There is a shortage of knives and forks: no fish knives and few spoons. We were too hungry to argue about it. The waiter talks what he thinks is English: "What do you want ?" is his rather brusque question. The waitresses have no English at all. After dinner we had a rest and then a stroll up our Boulevard Sebastopol and another to the Boulevard Magenta, and so back home at 11pm.

They say Paris wakens up at 10 o'clock. I don't know outdoor cafes and tables are the order - and frankly, they are not interesting to me. I suppose the dance and the music halls are lively : we may go out for some music in the next few days. So far Paris is a scream: in many ways its a city not unlike Liverpool or London but others are quaint enough.

Sunday night.

We have had a quiet but interesting day. The hotel seems to be entirely "tours" Frames, George Lunns, and others. We have a special Dean and Dawson table - which keeps us together. We are a mixed lot, and one couple, elders, are not pleasant neighbours. They sit at my elbow, and she grumbles at the cost and scarcity of things. Therefore I hit back, gently , but firmly. "Last night's dinner was good enough for me: coffee, rolls and butter is not a poor breakfast; I had a good time crossing, and really I'm not out to grumble if the hotel exchange is about 80 francs and the Bourse is 83 to the pound."

The breakfast was not bad, coffee, rolls and butter - not bad except that the salt was missing. French sugar is coffee proof, but I'm not troubling. Afterwards we strolled down to Notre Dame, visiting a bird market on the way. Annie was greatly struck by the dinky little lemon parrots about the size of sparrows, and some tiny unknown bits of feather which a fighting bumblebee would chase a field away.

Notre dame is magnificent.

It is everything one expected, and more. We attended a goodly portion of a service. The music was lovely. The stained glass is also very fine. The interior is quite gloomy, and there are scores of chapels. Outside, the island is full of glorious buildings, but we didn't go to the Palais de Justice, the Sainte Chapelle, though we passed the Hotel Dieu, a great and dull looking hospital. I prefer Liverpool Royal Infirmary for the stone is dark grey. The marvel is that smoke is so rare here: the coal is about the darkest stuff imaginable, and a railway train trails a dust cloud along.

We were moving leisurely. On the quays there was a mighty fishing industry. Now the Parisian fisherman is industrious. He brings the day's grub - maybe his wife, his uncle and a sporting neighbour. They use bamboo poles, few have reels, and their bait is the common gentle or meal grub. Annie complains that his grubs are lively : horrid. It's a mighty industry. I was leaning over the quay edge, and a chap passed me. He smiled, waved his hand, and spluttered out a few words. I didn't understand, though afterwards I recollected the first sound, "Jamais." Annie says this means "always."

The fisherman is now Paris Jammy: he is a sign of eternal quiescence. There is no change. Every Sunday comes he to the Seine; every Sunday it belches up little bubbles of bilge gas. Every Sunday the fish decide to stay at home. If my residence was in the scented, muddy river, I should give Jammy a fish now and then: but I wouldn't eat him. No, not I.

I watched Jammy for a few hours. He never seemed likely to catch a fish, and I think the crowd would have been as disappointed as he if he had got a real bite. No one asks the Paris angler "Cop'd owt?" It would be rude indeed. He is a symbol of "always"; he never finishes his job.

Our ramble along the quays brought us finally to the Louvres: this is the great Arts Museum. We did not enter but walked in the gardens, admiring the statuary. Here is one Arc de Triomphe, and look through it you see a long avenue: Gardens des Tuileries, chiefly a wood of dwarf chestnuts, then Champs Elysses - the Elysian Fields, ("when a good American dies he goes to Paris"). To me, the trees are sunburned and poor. It has been a dry year. At the top of the Champs Elysses, probably a couple of miles off is the great Arch of Triumph where France's unknown soldier is buried.

We walked , or rather adventured through the north gate of the Louvre, and walked down the Rue de Rivoli. The French chauffeur is an idiot, no less. I've spent hours in .him. On the open street he twists and cuts in everywhere; he accelerates for a corner, uses a dirty pip squeak as a horn; and pulls his car up, all - standing and engine roaring, with the brake. A merciful man is merciful to his engine : these men are not.

Our route home was along the Rue de Rousli with its Magazins du Louvres and other big department stores. They were closed of course, but the windows gave us a humorous.

At this point 9am Monday, my pen has given out. Annie and I went across the street and bought a new bottle for 60 centimes - penny size.

Lunch was quite substantial. It is the best meal here, with a decent cutlet of mutton. The fish was rather tough - it must have come from elsewhere than Dieppe, where pisces have to work hard for a living and not batten on Channel lunches.

There are many things I want to tell you, but the pen is too slow. I'll keep to the routine if possible. You will be glad to know that French food suits me. I'm too excited I guess to bother about my tummy, and there has been no trouble at all since the crossing. My unpleasant neighbour thinks it funny to point to the breakfast roll and say, "this is ham" and to the butter, "Pass the eggs." I bet they are "showin their browtins up." Anyway, let's be thankful. They keep me from grousing as I should have to agree with them even for once in a while.

12.15 Monday Your letter received. I knew Devon would catch you: it's glorious. Look out for a cottage.

Our Sunday afternoon ramble was prime. I asked the Irishman, (name unknown) to join us, and we wandered off to the Louvre. We did the usual ramble: it's the British Museum sandwiched with the National Gallery, with a top layer of South Kensington. We saw exactly 123.456789 acres of stained canvas, and then adjourned to the Louvres gardens for a seat and smoke. The finest thing we saw was Leonardo da Vinci's famous Mona Lisa - the smiley lady - which took him years to paint. He got more on two square feet of canvas than the others got on an acre. She's a smiling demon, and about 20 years ago, she was pinched from her frame. She was recovered in Florence, Italy in 1913. Here she rests until I decide to move her out. "The Raft of the Medusa" and "Justice and Vengeance Pursuing Crime" are horrible yet fine.

Paris is a city of statues. The most recent is a figure Paris 1914-8 - a defiant woman waiting, with sword sheathed in her hands, most defiant and determined. The face is hard and composed, with none of France's traditional excitement. "They may come but-." We strolled in the Louvre gardens with their lovely parades, through the chestnut groves of the Tuileries, where concerts are common. Across the Place de la Concorde with the twin sister of Cleopatra's Needle and the best Niagara of a fountain I have ever seen. It was lovely. Jean could have hired a little yacht and sailed it among those of the French children in the base of the fountain.

Then we walked a couple of miles up the Champs des Elysses, finishing at the top of the hill with the Arc de Triomphe d'Etoile. Here lies the French "unknown warrior." The Arc is a wonderful view point. Twelve avenues strike out from it, regular as wheel spokes, with trees down each side of the road.

By this time, we were thinking of being tired, so Annie and I and the Irish man turned down Avenue Marceau towards the river. On the way we sat down in a street cafe and

Eglise de la Madeleine

Trocadero from under the Eiffel Tower.

Annie ordered three cups of coffee. They came in glasses, with milk and little crisp rolls of bread, quite welcome fare, and we sat quite twenty minutes over it. Cost 50 centimes a cup, and something for bread.

The quays of Paris are quaint places. Down about Notre Dame we watched the folk washing clothes in the stream. The shirts were dark grey before they went in. I doubt whether they were two shades different, dark or light afterwards. In "Three Men in a Boat" Harris washed his pants and cleared the Thames. Half a dozen families did not clear the Seine.

A swimming race was in progress, ("through Paris"): we saw the last two competitors only. The ferry boats move quickly along the narrow stream, and yesterday were crowded. We arrived at the hotel at 6.50pm, ready for dinner.

After a long rest, Annie wanted to go out again. At 10 o clock it was quite warm, and we strolled out hatless to the boulevard Sebastopol. Later we sat down on a cafe seat, and had chocolate - 2 cups 1 franc (3d). Annie nearly killed the waiter by presenting him with the change out of a 2 fr. bit. It gave him a pain, and he bent nearly double.

We are making a sensation in Paris. Everyone sits up to look at me - or is at Annie. Her hair is much discussed. The Parisian ladies have straw - gold thatch, but never a head of hard red auburn. Footwear & clothes in Paris, as worn by the ordinary people are cheap and poor. There is nothing of the well dressed in this quarter. We got home about 11pm. Our cafe was on the corner not forty yards away.

Monday 5.30pm.

Today we have been motor jaunting. In the morning we went down to the Louvre, across the bridge to the Isle de la Cite, saw Notre Dame, and had a run through the East End. On the Isle, we saw the prison where Marie Antoinette was incarcerated, and from which she passed for trial. This was in the Palais de Justice. Notre Dame came next: we saw more almost yesterday on our own, though the guide did his best. Then we saw the site of the Bastille, and the builder's yard where the guillotine struck off the heads of aristocrats, men & women by the score in the Revolutionary days. The history of Paris is full of these tempests: she had five bust ups in a century, and keeps relics of the lot. Her guides spill out dates like beans, and then have a bit of wind for the Louis genre. There were umpteen of them on the throne. Some good ones, but one can't locate them. The guides jumble them most joyfully.

Then after prison, church and guillotine they playfully carried us on to the cemetery. Pere la chaise, to see the heavy weight tombs. The place is solid with blocks of stone. Presidents of France are side by side with humbler folk. I felt jolly enough to sing, and to recite Mark Twain's version of Abelard & Heloise when the guide was out of hearing. But there is a magnificent view from the higher walks, with the cupola of the Invalides and the twin towers of Notre Dame. By the way, Victor Hugo is a hero equal to Napoleon 1. The guide told me he was exiled to Jersey - a - Guernsey. I said, "Not much; he bolted." Victor's little jaunt would have been in a different direction if Paris had pleased itself.

One church was shown us where a shell from "Big Bertha" dropped on Good Friday 1918, killing 75 and injuring 90, chiefly women and children. The guide got lockjaw so I explained a bit for him: he wasn't too grateful, and regarded me with suspicion. He guessed - wrongly - that I'd been there before. It didn't matter much really; we had a new guide after lunch, an old stager who pathetically believes he knows Welsh because at Easter he shows Cardiff & Swansea football teams a bit of Paris. I didn't enlighten him too far. Anyway he understands my type of English, and rather marvels at my "get there" speeches. I'm afraid he'll try them on somebody else and get injured. He too disbelieves in the bottom of his heart my yarn about not knowing Paris. I chip him too promptly to be as ignorant as a good boy should be. Anyway the old fellow had 53 months 22 days of war, and 40 months in the firing line. I forgive him his omissions, if he didn't get a commission on our two lemonades, 3 francs. Wine is cheaper here than mineral water. Henceforth we have either coffee or chocolate. Marsh was right about the guides; the one this morning was very careful to say nothing outside his "lesson."

After this "dissertation" let me return to our afternoon in Paris. We drove out past the Louvres and into the Place de la Concorde, where the old boy pointed out the various monuments including the eight statuary groups presented to Paris by her sister cities. The Strasbourg monument used to be hung with black from 1870 to 1918 when she returned from Germany to France.

Then up the Champs Elysses, shortly turning over the Czar Alexander bridge to the Hotel des Invalides, which is partly military HQ, museum and hospital. Here we saw the railway carriage in which the Armistice was signed, and in another gallery, one of the taxis which saved Paris in August 1914. Gallieni whirled an army round Paris in all sorts of motors, and hurled them across von Kluck's flanks. The von he clucked, but he couldn't write the Kaiser about this umpteenth army driving across his rear lines, and so he halted. The result was the battle of the Marne, and the first German retreat. Our old guide loved me because I took my hat off to this good relic. I got him to translate the old taxi driver's reply to Gallieni's order. It came to this colloquially; "Where you says Boss, there we goes." It wasn't classic French he agreed, so I haven't used classic English.

After that we went past the Eiffel Tower, a big hoist of girders stuck on end. My unpleasant friend is going to the top; therefore I'm not. I can see my Paris from other places, and will do so. Its only cussedness to say such things, but I've no time for Eiffel, until its made of lemonade.

Now Paris is a queer place. It lives on revenge. If Foch had known that the Allies would have gone to Berlin, the Armistice would never have been signed at all. He would have driven the Huns home, and further. The hate crops up everywhere. The guide faltered when he showed us in the Place de la Concorde when in 1870 the Germans camped after the surrender of Paris. He was coldly furious when he recalled at the Arc de Triomphe d'etoile (Star) that Bismarck proposed to march the Germans through Napoleon's sacred gap. Paris defied the Germans and barricaded the thoroughfare, and Bismarck dared not insist. He might have shattered and burnt Paris, but his army would have melted to pieces. The day's camp on the Place de la Concorde cost him hundreds of men murdered by the furious French. The Germans wanted to change the name of a bridge, d'Jena, but again the French declined. No power under heaven could drive them out of their determination. Keep the name of a grand victory they would, and did: Bismarck was defeated again.

Paris has another joyful habit. Every ten years or so she plans an international exhibition, builds some mighty palaces, and makes money out of the venture. At the end she makes herself a birthday present of the buildings. Her museums of fine arts, the Eiffel Tower, the Trocadero across the river are merely some of the exhibition buildings Paris has annexed. She is now preparing a show for two years hence, which will mean some more permanent buildings for the city. The best part of the Trocadero is a concert hall for 6000 persons! On the way to the Troc. we passed the house of Georges Clemenceau, the "Tiger" who was the war time President of France. A fierce old partisan, bitter of pen and tongue, he was patriot enough to sink all in the hour of crisis to save his country.

I took my hat off to his house; and might have taken off my socks if he had appeared in the roadway. Alas! that Paris is empty: the notables are all away on holiday, and there's only four million other folk to welcome us.

Again the Arc d' Triomphe: I spell it differently every time. Our guide gave us the usual items about the Unknown Soldier. The epitaph is very simple, but lacks the poetic genius which inscribed the words of our own British tomb in Westminster Abbey. There is a brass slab to commemorate the return of Alsace - Lorraine in 1918, and a glorious view.

Now let me baste our bus driver. He edged us to eternity again and again, cutting in, bouncing out, jamming and jarring - granny needle wasn't in it, and I wanted him to charge a hand cart and be done with it. The old guide grinned when I appealed that he should crash into something cheap and get it over with. But he wouldn't. The assistance of a thousand racing idiots didn't get us one crash. At last we nearly did it; another coat of paint and there would have been a collision. We were ambling down an avenue at the usual forty knots when our friend and another made a race for it. We wanted to reach a park, and the other idiot merely wanted to cross our bows. Well, he did it, but we forced him to do a two step on three wheels and a turn of the binnacle to the southwest before he got past. The other fellow and his passengers cussed gaily; there was no bite about it. Our Jehu jeered and told 'em what he'd do next time. I hope he does it - if I'm not aboard. These car drivers will be the end of me. I shall take the laughing sickness, or learn to shout Welsh, or suffer some other damage before I get home. Some of the taxis are curious: they look as though some one had sat on them. Not one in six had a straight frame, and most travel with a list which keeps the driver turning hard aport in the straight boulevard. How can a fellow guess the way they are coming.

The police here are gorgeous. With a hefty boxwood baton, an occasional pointsman holds up the stream of traffic while the drivers light cigarettes and howl abuse as to whose wheels are you jamming, and why don't you keep your tail out of my eye. They do more yelling over the right to turn a corner than would keep Everton Football Club going for the first ten minutes of a cup tie.

But I've lost my Paris trip. We saw monuments of two criminals: I had been taught that Hansard was an architect, now I find that he was a No.1 Tax dodger. Windows in

France were, and are, taxed and the poor Surveyor or Assessment Committee was bullied into a statement that of course roof lights were ventilators, and therefore exempt. Hansard did 'em brown, and I love him. He built houses with no walls and fine long roofs through which he pushed his windows. No taxes, and a statue: thus is honest effort rewarded. I've forgotten who was the other criminal - oh yes! This one collected the following recipe;

```
cocoa ........................... 1 centime
sugar ........................... 1
cocoa butter ................ 1
silver paper ................. 1
painted cover.............. 1
```

and sold it as chocolate Menier for one franc. France hasn't decided what to do with him, so he lives in the most gorgeous of houses overlooking a fine public park. He has the pick of Paris whilst Clemenceau lives in a second rate street in a big block of "mansions." C. saved France, but M. got the money.

Where's that Paris trip? Bouncing on at forty knots, chasing cyclists and pedestrians, and almost shooing the police into the gutter. Finally we reached the Madeleine, the society church of Paris. A German bomb fell in the square and knocked the head off a figure in a niche. Does the Frenchman repair? Not a jot. He gets a skein of barbed wire, and stilettos, a few laurel leaves or thorns on it, and with an artistic bend, two nails and a bit of string, your headless statue is adorned with either a laurel wreath or, a crown of thorns. I'm tired of grisly details, but the guide isn't. H has the wheel turned 180 points astern, and we are on the front of the church. Enter, no, not yet. On these steps 200 of the Communists or other revolutionary heathen were shot down by the army. The Communists had barricaded the avenue solid, but the soldiers didn't play the game. They charged through a side street, and swept the front of the Madeleine with bullets. The other fellows tried to rush into the church, but the leaden hail swept them off, though one or two died on the threshold.

La Madeleine is a church without windows. They are three or four bulls' eyes in the roof, but candles are used in the chapels and on the altars. It's a stately gloomy place, and people go there to get married. I've nothing more to write about it. And our next run was back here to the hotel (5.25). A few minutes later, Annie was off again, and we made a short curve of the boulevard. My nose smelt printers' ink and we found the Fleet Street of Paris with Le Matin giving a wireless selection on the bells. We came across an elaborate window display of a new washing powder which does without soap, scrubbing or water. At least, so I read the legend over a barking dog, some scurrying ducks, and a farmyard in working order.

It is now 1 a.m. I really think this holiday writing is a joy and I'm making a record for future use. We have seen much more than this. So far I have written no articles.

After dinner we went to our corner cafe, but although it is 30 yards away, the journey took up 2 hours. We turned in the wrong direction, and followed the Boulevard east and south to the square of the old Bastille. Then down the street to Henry IV Bridge and to

The Forgotten Man of Lakeland

the island on which Notre Dame stands. I guessed a moonlight view: it wasn't very bright, but it was effective enough. The great towers, buttresses and spires shone ghostly above the street lights. It was a dream - but I want to see it in a full moon of winter with a sky like steel, and the stars mere twinkling lamp posts up to heaven. Then home across the bridge, by the silent river, past our great landmark, the tower of St. Jaques at the foot of the boulevard Sebastopol. There is a big figure on a turret: I have christened it after my fisherman Jamais - Jammy - always there. The little waiter gave us two excellent coffees at 40 centimes each, & got a 20 centime, (3/5th of a penny) tip, which pleased him.

Tuesday 12 noon.

This is a free day, and gloriously hot. I don't wonder that Paris is a city of trees. They need all the possible shade in summer. The parks are all short trees, in avenues, with plots of grass and flower gardens patched in. You can go anywhere among the trees, they are chiefly small chestnuts, and keep cool. This morning we were driven from the streets into the garden of the Luxemburg, and were glad of the shade. We have dodged home to lunch and to shed a few heavy clothes.

I can't recall the route. We have been in the Latin quarter south of the river all morning; saw the University's outer walls and the outside of the Pantheon, went inside St. Etienne, which is really the church of St. Genevieve, and then into the gardens of the Luxembourg. In the old palace the Senate of France meets. It's about as interesting outside the palace at Kew : they don't let the public in to see what it's like inside. Perhaps it's just as well. Statuary by the way, is a tiring study; one can easily pass a stone or a bronze group here. It's only one of many. The Luxembourg gardens are first rate but not so good as those of the Tuileries say about equal to the best of Sefton Park, with belting big tree ferns growing in the open. We had a hot walk back, and (I'm writing this at 6pm) there was a sequel. The Blue Guide says that the Boulevard Sebastopol is ugly; we agree most thoroughly, especially after some hot walks along it. This morning we dodged down a parallel side street, the Rue St. Martin, and got into the fruit hawkers' quarter, with yards as narrow almost as Kendal, and a smell of fruit both ripe and over ripe, but not unpleasant. We dived in and out of the shadowy cuts, and had an interesting time. We finished up at the Hotel de Ville, by the side of the river, and saw a flower market where on Sunday the birds were on view. It was bright and gay of colour.

To come to the sensations of this afternoon. I asked the quiet Irishman to join us if he wished: he came. First that mile of the Boulevard Sebastopol, and then we took steamer down river to Auteuil as the coolest way of seeing the river. The Seine isn't the Mersey, but it did its best. The sea breezes don't travel a hundred miles inland, but we had the shadow of many bridges. The Eiffel Tower and the Trocadero were visible, but the banks are high, and there is no great outlook. The cost was about 2d apiece.

We didn't stay at Auteuil, but boarded a return steamer at once. The Irishman insisted on paying this time. Dodging under bridges we came right upstream and got a lovely view of Notre Dame from the water, getting off at the first pier beyond. It was now 3pm and furnace warm, so we joined in the Parisian industry of "watching other fellows work." The first was a professional strong man, who juggled with an iron ball on his head, climbing a rickety step bridge, and throwing 1 kilo iron bits which he had picked up with a palm grip. Annie translated his invitation to us - one and all - to do likewise: it was too hot so we strolled away. After helping in washing sundry dogs, we joined a committee on coal heaving, and watched the men fill bags of the dirtiest abomination Paris calls coal, and carry them ashore up a ladder, over a gangway and dump into a cart. How hard worked we felt.

But that was nothing to the next feat. From the dizzy heights of a footbridge, we entered upon "le sport." Jammies in plenty were fishing. These were just the fag ends of the umpteenth million, no more. Perched on the quays, the sloping bank, on our bridge, and in punts, the brotherhood - aye and sisterhood too - of hope fed more and more grubs to the tigers, lions, wolves bears and alligators of the river. Then voila, v'las, v'a -z, if you will, down , down, down, z -a, went the quill float to upright, then the water rose over the red stained top, and Jammy struck. Heavens, how the rod bends, the line twitches, the angler sweats and raves. Cold tears ran down my back. But Jammy the inexorable triumphed. Up from the deeps came a silver shadow, a twitching, fighting, sorrowing giant of the river. There is a flurry on the water, a pitiful lashing of a tail. One moment that king of fishes flashed in midair, the angler reached, disgorged the hook, and deftly tossed his captive into a tank on his boat.

Ask me now: are there fish in the Seine? Still I know not, but I know that at 4 pm this afternoon, there was; at 4.1, the question went back to the realm of doubt. Jamais - always is justified. But alas that I know not what happened to my prince of fishes. Tonight at dinner I shall salute him as a sardine, or perchance he may come for final entombment among my whitebait. Let him but be cleansed of Seine muck and ooze, and I shall eat of him in gladness and sorrow for was he not a noble fish, even before I made his first acquaintance.

After our desperate capture, we felt, like Izaak Walton, that it was home for a drink, but ours was not old bottled ale, cooled in a spring. It was 3 coffees at the nearest cafe. We had surely earned a great reward, and we made a desperate resolve that at 5.15pm we should take another drink - which we did. Annie's French held out.

We walked across the Ile de la Citie behind Notre dame, and then across to the Latin Quarter visiting a church on the way up the hill to the Pantheon. By this time, Annie has become adept at ordering drinks, and the game went on all right. We sat awhile by the Boulevard St. Michael - the Boul Mick of the students and got lazy. Then Annie bought me a collar stud for 65 centimes - about 2 1/2d - and at the prospect of the long trounce up the Boul. Sebastopol with faithful Jammy squinting down at us from his tower I got tired, oh so tired. The sight of a gang using hydraulic bars to break up the street made me quite industrious - how could my energy be worked off. Ah! There I saw a map of the Metro Underground. I drew mental lines on it and decided that Heamuer. Sebastopol was 4 stations away. Therefore down the steps out of the heat and glare, with Annie working off a heap of change and French to get the tickets. Down again to the station, and then a train came in. The Metro are humorists: the second class carriage has a couple of boxes at either end in which the lucky ones get seats. The rest grab or lean against brass rods from roof to floor. The going isn't bad; it seems slower than the London tubes, but may not be. Anyway, my gamble on the French map came off. The fourth station was about a block away from home. We dodged those miles of street tramping, for which relief we were duly thankful.

Now to get back to my old, old bugbear the streets of Paris. A million English folk have been to Paris - ah yes, and they do agree that the glory of that city is in its streets - wide boulevards radiating like spokes from a wheel, and a rough circle of roadway right round the city. Now comes one man with an eye and a memory. There will be the deuce to pay for this.

Paris as an ideal street city is a delusion: the actuality is not a dream, it's a night mare. I grant there are boulevards as stated, but two thirds of Paris traffic drives, unregulated, through streets full of twists and bad corners, at a fiendish speed. I don't dare to ask about casualties, but they run a double bus service right and left down streets twelve feet narrower than Bold Street. Tram cars scream round corners, and our 18 seater charabanc had to spin like a turn table round some of the cross streets. The area inside the Bastille to Madeleins Boulevard is worse than the city of London. There are more blind corners than open ones. It's worse in the Latin Quarter south of the river where the streets are steep as well as twisty. No! Paris as a great road city amounts to Bunkum and I'm not going to forget it.

As for maintenance, the tram line down the Boulevard Sebastopol is the limit. It would disgrace the most backward village in Britain. Elsewhere, the same applies. Take the centre of Paris, and you will find pavements set by the Romans and worn, picked up and turned over by every generation since.

After dinner this evening I am writing in my room, coat off and sweating. At 11pm Annie and I wandered off towards the Madeleine church, then followed the great boulevard Capucines, & to home again. The street cafes were crowded; there was a certain amount of music; a flashing of sky; signs, and plenty of well lit shop windows. Bare arms and necks and backs flashed everywhere. Present day fashion here is for ankle dresses, which makes up for deficiencies elsewhere. Three out of four women rouge their cheeks, four out of three are carmine with lip salve. You get a great flash of it at fifty yards away, and it looks worse every step you get nearer. Its a sorry show, but when Mary Jane rubs her top note with black lead round her eyes, and boot polish for her eye brows, the offence screams for shadow against the sunshine. I don't guess what happens about their hair: it's lack lustre stuff. Like hay or straw at best. Parisian ankles - my hat. I have yet to see any worth noting. The present fashion decrees loose dress round the waist here, and no corsets. In a motor the women look fairly good but stout. When they wobble on two poor feet along the street, I've got to laugh.

This suggests that Paris has the loveliest big cats I've yet seen. On average a fellow would clear Salisbury Road in ten minutes. They know that they are the elite, and compose themselves in the haughtiest fashion. The dogs are either alsatian, mongrels or smooth terriers : good in condition and either muzzled or on leash. We saw a poodle tonight : the first, and an unclipped chap at that.

We had a vanilla ice each at a street cafe. Annie pronounced the stuff good: the bill was 500 centimes or 1s 3d. And the waiter had to run round to a branch shop for the ices at that. The cafe man doesn't say "I'll order it for you." He goes and gets it. These men get no wages and yet some die rich. In our hotel, no tips are given, but 10 per cent is added to the bill for "service." At the end of our ramble we called on friend Alphonse for 2 cups of black coffee, 40 centimes and made his heart rejoice over the other 20 centimes - 3/5ths of a penny. Every customer gives the waiter from 5 centimes upwards, and so come the millions - in time. The waiter may die first, or go bankrupt as the owner of the old cafe.

Wednesday - lunch.

In one way, there is little to write about this morning's outing. The Louvre is too vast to mention. We walked through a mere corner where the greatest paintings and sculptures were on show. Venus de Milo is a splendid bit of work, and one of the world's marvels even in its broken state. No one can overlook it as a masterpiece. The badly broken "Winged Victory" is another marvel. The French "Crown Jewels" are scarcely worth a song: there are few, and of these, the Crown rebuilt for Napoleon 1 is about the most important. About 40 years ago, the Republic sold off most of the jewels: it was very hard up. Millet's "Reaper" is a good painting: the "Angelus" is part of a special collection which we did not see. From the rest there was a fine show of 16-18th century work: all the great Dutch, Flemish, Italian, Spanish and French artists are shown here. There are acres and acres of saints and sinners with here and there a beautiful bit of still life - flowers, fruit or fish. One gentleman had painted a butcher's carcase, even to its greasy appearance. English artists were scarcely visible at all: Constable, Gainsborough, Romney, etc. were not represented in the Grand galleries that we saw: they may have place elsewhere. Compared with the National Gallery of London, the Louvre has few of the early crude painters. The point of many historical portraits is lost unless you are very sure of your recollection of French kings and events. They had about 18 Louises, 10 Charleses, and some miscellaneous folks like Cloves, Charlemagne, and the saints. So far, we have seen no relic of the Roman occupation or early Gaul: there must be museums elsewhere. I should think there are 60 at least, galleries and museums in the city.

Their museums here have a brighter and less frowsty appearance than our British stone and paint shows. They smell less of carbolic below, and dust above. Perhaps the air of Paris is a bit cleaner: it's much less smoky.

After lunch we had a drive round the city, to the Hotel des Invalides again, where this time we saw the tomb of Napoleon. You must bow to him whether you like him or not. The architect arranged a balustrade over which you must bend your tummy to look to see the great sarcophagus of Russian stone which contains his body. The French are proud of showing his twelve victories: they say nothing about Waterloo, and Wellington's great campaign which rolled them out of Spain and Portugal is never mentioned. They show few sea battles, and never show a sign of Nelson. There are things in history of which a Briton does not wish to be reminded.

After the Hotel des Invalides, a trip to the Pantheon, where good Frenchmen are buried, and then to the church of St. Etienne which we saw on Tuesday. By this time, our keenness for sightseeing had been worn off. Our party has been unfortunate with diarrhoea: one man had a bad time on Monday and Tuesday: a lady was in bed all day

Wednesday: I had a touch yesterday, but it soon passed off. It is variously blamed on the Paris water, and on the oil in our hors d'oeuvres and salads. I blamed some bitter lemonade at lunch yesterday, but Annie was not affected at all.

After the drive, Annie and I rambled down the side streets on a shopping expedition. We sat down in a cafe, and had 2 cups au naturel - 35 centimes each. Then we donated to the waiter the remainder of the franc and gave him stomach ache. Then on to the shop where Annie purchased something. I spoke in English to her, and the lady replied. She could talk very well, and we gave her a few minutes practice. She was a good deal better than Philip our guide who mixes his English numbers and tenses remarkably well. Annie has bought me one collar stud and 2 pairs of cuff links, all work done by her own French.

After dinner, my stomach not behaving well, we sat in the hotel for an hour or so ; then walked, I in slippers, to see Alphonse at the corner. He could not give us milk with our coffee and was very sorry about it. A short turn up the boulevard ended the day. At 11 I was in my room, and felt tired enough without real hard work.

Thursday 6pm.

This morning I received your postcard with 30 centimes to pay. The "Fell Race" article must now wait for another year. I have written nothing from here but this diary. I'm not sure I have the material as yet. There will be enough time when I get home. I am using odd gaps of time at the hotel for writing, instead of sitting in the lounge, swopping English lies.

This morning I awoke at 8.15, the latest this week. I suppose my little trouble caused me to sleep harder. Another lady of our party has been devastated, but she kept up all right at Versailles. I haven't caught her drinking water. White wine (graves) is 3fr 50,(10 1/2d) for a small bottle, 6frs. for a big one, so wine is constantly drunk at table. The French beer is also popular: It is light straw colour, and hardly at all intoxicating. Champagne is as low as 3s.6d a bottle. Wine is cheaper than lemonade: and the lemonade, even at 3fr a bottle is not nice stuff. Some lemonade we got at the Trocadero was the limit. It has no flavour and little aeration about it. Coffee seems to be the best general drink. We spotted a place last night where it is 10 centimes a time. It varies greatly in both price and quality. Tea in the hotel is more usable than elsewhere. Today I have discovered, with the aid of Philip our guide, a drink called St. Galmier, a natural mineral water, not unlike our soda water. It is quite good to drink, and varies in price. At Versailles it cost us 2frs 50; at dinner in the hotel it was 3 frs. Philip said it varies from about 1 fr 50 to 3frs, according to the hotel. There are places where it costs 1 fr. (3d) a bottle: the wholesale price is about 50 centimes. There are fortunes made in the hotel trade. Our proprietors claim to be Swiss: they certainly don't look very French. Tonight, the boss has had into dinner an enormous fat fellow. His waistcoat would make me a suit, and his jacket would do for a sleeping bag, and pants.

Now, whom do I abuse today. Have I admired the painted beauties of Paris: I think so. Their lips are danger signals. Let me deal with their feet. So ho, for my childish hope that here they would know what to put in their stockings. Instead they are stack heeled, slab footed folk, shovelling along like tramps. English folk would stare at the average

French waddle and hobble. I don't know about the fashionable except that in the great shops the gear is on different models to England. They allow for a long, not a high instep: there are English branches here showing this as well as the British type.

Now at last having condemned the ladies, root and branch - the average Frenchman isn't worth a line - let's move to Versailles. We had yesterday's amiable driver - Weary Willie; dashing Albert of Monday was out with another party. I nearly lost some money over him. I backed him to ditch one of Frames' cars if we saw the gang on the road to Versailles. He certainly drove with some verve, and there was a sporting chance of making them eat our dust if he arrived. He didn't; anyway, we didn't see the Frames on the road so it doesn't matter.

There was the usual scrum in the narrow streets; twice we went through before standing lorries with an inch, and some loud and radiant language to spare. Up the Champs Elysees to the Arc de Triomphe l'Etoile, we ripped a bit, and on the Avenue de Boulogne Weary Willie let her rip. Jemima was a 22 seater, and we bounced her up to 43 kilometres an hour (about 25 miles) on a long incline. After that, I ceased to plead with Philip to smash up a baker's cart: I always wanted something soft and cheap.

The Bois de Boulogne is a big wood, chiefly chestnuts, with the city forts; it is a glorious place, and no doubt in the spring society "walks along the Bois Boulogne, with an independent air." There weren't many folk about, and the roads were patchy. Every mile or two in Paris they are doing street jobs, and to all appearance the work is done by the day. One task is to pull down the old forts, and put parks and gardens in their places. My hat! But the Frenchman worships trees. He wants his flower parkeries out of a dress maker's box, but he is willing to let "le Bon Dieu" have a partnership in the trees. You never saw a garden in corsets, with carmined lips, blackened eye brows, and shaded eye lids! We have. They can keep them in Paris. They can't grow grass: it's a dingy thatch like that of a fair haired French lady. The bois or forests are really prime: they are tame and trim; they haven't the tang and majesty of Scots fir; nor have they the glorious oaks of England. The French idea of an oak tree is a tabby not a fighting tree at all.

But what about Versailles: it's Chatsworth and more for size; its Chatsworth and more for gardens; but it isn't a patch on Chatsworth for the lovely Derwent. The French idea of water is a tank where artificial cataracts and fountains are easily managed. But the great gardeners have done their best, and after Philip's infernal lectures on Louis the umpteenth who built this and that, its a relief to know that his name don't rest on the flowers anyway. Versailles is everything one can say, and more. It has its history and a hundred thousand pictures. I counted them - or at least I began to count 'em, and if they had stopped at fifteen (when I did) I should not have had to accept an estimate (probably Philip's) Poor chap, in an hour and a half of paint, polish and furniture he utterly mislaid his English grammar: some of our folk mislaid their legs. "Are you coming back this way?"

"Yes."

"Then you'll find us here."

We did: the plush seats weren't bad. I signed a post card on the spot in the Hall of Mirrors where the Peace Treaty was signed. My writing was about as poor as the Treaty.

There wasn't peace between me and Philip. He pointed out the painted warriorship of Louis the umpteenth. I said they didn't win battles that way. Louis always took three fourths of the paint; he left the rest for the powder. He was far too big to be true, and I said so. "No Philip; I don't mind you forgetting Trafalgar and Waterloo, and I know that you never heard of Crecy and Agincourt, but this thing reminds me of a yarn in Scotland. There's a good stream there, and one village claims to catch 20lb salmon. On the other side, the villagers catch 30lb "fish." A "natural" was informed of this: "na, na, they're no bigger fush; there's bigger lee-ars across the watter."

We saw bathrooms and dressing rooms, drawing , dining and reception rooms. And rooms which had been used for the lot. Philip at last led us into a long salon, and I threatened him: "Philip if you say this has been a dining room, I'm going to put you down, and sit on your head." So Philip quailed; he said that Louis the Umpty - umpth had foreseen this little trouble. He wanted a gallery for 12 3/17ths acres of battle pictures, so he routed a boarding house of princes and poorer relations, and cut away the walls of tiny rooms to make a skating rink sort of place.

In another place, the artist had mis-guaged the size of an African battle picture. He provided 140 yards of spoilt canvas, and the room is only 100 yards long. Therefore they ran up another instalment on the west wall. It doesn't matter really. The picture tells one gorgeous truth. It shows the fleeing Rothschild, running away with the cash box while the soldiers are breaking up the enemy. France could, probably gain more by repainting a couple of yards of canvas than she will ever get from the Germans. There's enough wars to sicken one on the walls of Versailles: Kaiser Bill was going to govern from here like his grandfather did. The trick didn't work the second time: it was Bill, not France that got the noble order of the knock.

Now, let me say that this record of Versailles is entirely out of order. It came after lunch. Before that glorious feast we spent an hour or so in two other palaces - the Petit Trainon and the Trainon. The Royalties went to the first when they wanted to use back kitchen manners, and eat fish and chips with their fingers. Marie Antoinette was too big for a dolls house: she had a model Swiss village with mill and dairy and real cows which she used to milk. This toy is almost in ruins now: the French folk who beheaded her has done a good deal to keep her memory alive. The Royal rooms in the Petit Trainon are quite simple: it's merely a 16 room villa done up simply, but regardless of cost. Marie told the market woman who had stormed Versailles about the scarcity of bread, "Why don't you eat cake! The shops are full of it."

The Tranon is a small palace: a sort of Imperial waiting room for Versailles. It's another lovely place in the same great park. All are national museums today. At lunch, one of our party scared the waiter. The French word for fish is poisson, and he tried to pronounce it. He got "poison" instead, rather an alarming statement. Weary Willie nearly crashed us going home, but alas! it wasn't his fault. At Petit Trainon there are carp like whales, and water lilies with a pink tinge. The palaces are really grand.

The Forgotten Man of Lakeland

Friday 9am

Written in a third floor back of a Paris hotel, with a charming view of street noises and a cement wall. The place is certainly suitable for literary expression. Our place goes 7 up, and the lift does not come down with passengers. Writing is a disease - I want to do 100 folios on Paris. This is 71: it's less enervating than reading so called English and American newspapers.

We got in from Versailles about 5.30 last night, and strolled a little before dinner. The markets were closed. So that awaits our next trip. Our evening's entertainment was at a French cinema where the best seats at 2fr each. They started at 8.15 with Pathe pictures, and then proceeded to a Wild West gambling saloon love story with fighting, murder, a fire and trial by jury. Then we had a lecture on Shackleton's South Pole expedition: the clearest speaker I've heard. Annie said she got the drift of his remarks all right, but she didn't laugh at his jokes. The penguin scenes were very funny. Last we had American comic, a gymnast who did crazy tricks. The finish came at 11.25.

The audience was Alderson Rd English. Their behaviour was excellent. Of course, the young fools indulged in kissing but they do that as they stroll down the Boulevard de Sebastopol. It's very queer to see a couple check their stroll, and go through an osculatory exercise as calmly as if they were looking into a toffee shop window. No one takes the slightest notice. At half time there was a rattle like artillery. I wondered what had happened when I remembered the tip up seats. Certainly they have some spring in them. A little girl seated in front of us was swinging happily on hers all the time. She could only push it down by sitting on the front. The experience was worth while, and we were drinking Alphonse's coffee at 11.35 pm.

Friday 12.15

This morning we had a first go on the French trams. We wanted to get to the Bon Marche, but first of all to the Luxembourg. I worked out from the Blue Guide - which is absolutely worth its price and more - that 29 car would bring us near St. Sulpice. Annie used her best French ; we were travelling first class; it made the conductor blink, but he got us the right tickets, even demanding 10 centimes more in order to give me a franc change - 55 cs a seat. Near St. Sulpice, I got a line on the church, and was for getting out. A polite Frenchman asked "Where yew vant?" Like a thick head, I gave him our pronunciation, which he did not recognise. No wonder, either. He was a bit shaken, but "Vil you write it please? I cannot tell your words." Annie gave him the correct name; "Ah yes, next stop." I would have got out anyway because I knew where I was, but he was proud to help, even waving his hand and pointing in the right direction as the car went on.

The Luxembourg Museum was really fine. It's the work of living or recently deceased artists and sculptors. Why don't they have a literary museum as well? It would be worth while, wouldn't it: but what bladders of gas the heads of some authors would become. They're bad enough now. We had a good hour in the rooms : costing 1fr.

Our next point was the Magasins de Bon Marche, a famous shop where the assistants wouldn't blink if you fired a Royal salute in Gaelic. They're sound proof, I'm sure. By the way, I really doubt if anybody in Paris is without some English. Alphonse last night was pleased to observe "coffe wiv milk." He ran off before I could put him under the drain, or sit on his head. Most of our folk now patronise Alphonse for a last coffee, and he is getting quite happy. I saw him drinking wine with his supper roll one night.

Better than the Bon Mache was the open air market in the Rue de Rennes. Here the French housewife in black and white comes down to argue with any trader. The row rises to high heaven : the trader has a hectic time over a 5 centimes piece. One lady did give him a dissertation on highway robbery in artichokes - green fine fellows - but as he was pointing out at the same time the rise in the cost of fresh air, the argument did not get much forrader. She won however, hands down, gave him a pile of wee coppers, and departed. In vain did he plead for a sous for himself. Philip told me that the purple artichoke is eaten like marrow, but that it has very little taste. The vegetable marrow is scarcely grown in France, and the best potatoes come from Brittany. The purple is a climber, it is grown on walls. Plums are plentiful this year - the country folk will not be able to sell them. Potatoes are small owing to the drought, but they may recover a bit before October when the first frosts come.

After the Bon Marche, we caught a car home, travelling 2nd class for 30 cents. I told the man, "doos Roue Tobago" and he seemed to understand. I also held up two fingers, and the same fare applies to every stage within the city. So after all, it wasn't so remarkable. Some of the steerage passengers (who stand in the waist of the car and block the way to and from the other classes) had tickets at 15 centimes each. If a 100 centimes makes 3d, this must be less than 1/2d trip. Travel is cheap in Paris, especially on the trams. You travel 2 miles straight away, and 2 more up and down as an extra. And the up and down miles are ever so much more exciting than the merely forward move. A Liverpool tram conductor soon "gets his legs": in Paris he can't. I have said that all worn out taxis travel to Paris to shake out their last stages of existence. Equally all worn out pavements and battered rails go into the Paris tram service. You get more bounces out of your money than you can get out of a German mark. However, we arrived at our hotel for lunch by the aid of a foot policeman with a boxwood baton, and a mounted man who automatically blows a whistle and forgets all about the traffic until he wakes up again. It takes five men to manage the Turbago, Sebastopol corner. There is something delightful about the baton work. A British policeman's hand drops like a signal, and the motors begin to buzz. The Frenchman hurls his waiting battalions forward with a hefty wag of his stick: "Get a move on - hi, hi, hi. That's a pedestrian on the road at the Gare de l 'Est. Get him, sic." There is a viciousness about Paris motor driving which I cordially dislike. "French Self Taught" hasn't one tenth of the epithets I want to hurl at the blighters. I must take secondary education in the language. I've given them all the Welsh & Gaelic words I remember, and frankly, they're not satisfied. Martin Conway's sonorous rhyme takes too long to recite, but the tail end of it caught the last man in one of the motor queues. It seemed to lift his ears and certainly he tried to stop for more of it. I think I'll try it on Philip if he comes within reach again. To return to my police, the Frenchman awakening from forty winks during which the traffic has shuttled in all directions, rushes into the breach of the street, and threatens the least offensive driver with his club. Then having blocked the traffic, he proceeds to take somebody's name and number. If the whistle goes during the process, the traffic must wait until the job is finished. Then like a yelling flood the cars break bounds, and traffic is travelling in umpteen directions. The whistle toots, and the confusion goes on until my bobby finds another victim. Between the bad streets, worse cars, and worst speeds, a Paris boulevard crossing is a horror. A boiler factory in full swing is child's music to it.

Now let us descend to earth, and to lunch. This is from 12 - 1 and should be called day journey. The French do get a bit mixed up in their goings - out. A day's journey should begin at dawn. I nearly bought a Paris sign - Defence d'afficher- really. Stick No Bills. This is stuck up on churches, police stations, anywhere and everywhere - and even across the third storey of our building. The street corners are plastered with it. There is a yarn that an intelligent Scotchman visited Paris, and prior to leaving his hotel he copied down its name. As soon as he was lost, he showed the card to a policeman, who laughed. Sandy was most offended, and would have slain the whole force but an Englishman came up, and made the peace. I always thought that a tall yarn: it isn't. It's easy to copy Defence d'afficher than any other moral text in Paris.

Today there has been rain - at noon, a tremendous thunder shower a few seconds after we arrived at the hotel. In the afternoon several pelts. Today is the saint - day of St. Louis, King of France. Perhaps that accounts for the arrival of black in the streets. The dominant note of the place is black. We spent a few hours and all our spare cash at the Bazaar Hotel de Ville. We believed that the prices are lower than in England, but that's open to correction. The bigger things are quite as expensive as at home with the exception of furniture which seems extremely cheap. The French do not include a wash stand or a chest of drawers in a suite. Every bedroom is fitted with a proper lavatory basin with towel rail etc. I think stationery is very cheap. It's not worth carrying home however in one's luggage, being far too heavy. The pads which cost me 6d in pre war days are 1fr 45 here today.

During the showers, we have sheltered in various cafes. In one Annie's French failed a request for chocolate brought a flurry of words. So I wrote down the name on a scrap of paper. He explained, slowly, that they had no chocolate, so we had coffee. After a while we had a walk up to the top of our boulevard, and found there Gare de L'Est, station of the east - and then we dodged back on account of the rain. This afternoon, there has been a succession of brief showers, the first for months.

After dinner we had a short final stroll to the west of Sevastopol, a trip which ended in tragedy. The faithful Alphonse, of the Cafe Biard was having a night off. It cost them one franc. His deputy had to be content with 20 centimes only. Perhaps Alphonse's attempt at English last night has ended in lockjaw. However we shall never know. He was a worthy Frenchman, the height of three sous, but a first class brother. He must have covered miles - in a week. Tonight, I saw my first Parisian with uneven legs. He sat in our cafe a bit, and then he tried to negotiate the way home. He expressed a desire to join us, but there was nothing doing.

Now at 12.10 a.m. let me do a final weep over Paris. It's a gay city of one cafe multiplied by a thousand. In one they sing, in another they dance; in all they sit at little tables and drink multi -coloured liquors. Compared with the other members of the party, we have not seen much. The Rat Mort, the Moulin Rouge, and kindred resorts are not in our line. Nor have we tested many drinks and viands. The hotel plus casual coffees has been quite enough for us. But we have seen the real life of old Paris in a way they don't know. I at any rate am satisfied: gaiety of drinks, music, lights and women is not in my line. I think our impressions will be the clearer.

Therefore, I wave farewell through my window to Jammy with the fishing rod, to stone Jammy on his tower at the foot of our boulevard, to Tom, Dick and Harry who will merely motor over me in my dreams, and especially to Alphonse of the Cafe Biard. If his shadow grows much less he will entirely disappear.

Author's Note

While every attempt has been made to accurately transcribe William Palmer's hand written text, some minor discrepancies may occur.

The Pantheon, Paris.

Chapter 10 - THE BEGINNING OF THE END

William Palmer died on 26th December 1954. The cause of his death was due to the disease which was then known as Huntington's Chorea, but is now known as Huntington's Disease, frequently abbreviated to HD. Even today, most members of the general public have either never heard of the disease, or know very little about it. Many medical practitioners also find that contact with patients suffering from HD is outside the realms of their experience. In certain cases, diagnosis may be difficult as some of the symptoms are similar to those of Parkinson's Disease, Altzheimer's Disease or even Muscular Dystrophy. Many of the early warning signs of HD are those common to other ailments. Some show a little unsteadiness, jerky movements, moments of forgetfulness, or a tendency to be clumsy. Other symptoms manifest themselves in changed behaviour patterns of moodiness, unreasonable outbursts of anger, or depression. It is very easy for the relevance of these early signs to go unnoticed by relatives, and it is only when looking back, from a distance of a number of years, and with the benefit of hindsight, that many realise that all was not well for some time.

Most of those who suffer from HD exhibit rapid, uncontrolled movements of the limbs and head. Some, which may start off as minor twitchings, can develop into wild exaggerated movements of arms and legs Another form of the disease has the reverse effect. This is known as the Westphal Variant, in which limbs become progressively more rigid.

In 1872, five years before William Palmer was born, an American doctor called George Huntingdon, was able, after many years of research, to publish his findings on this illness that attacks the brain, and causes a deterioration of the body's central nervous system.

This condition, which took his name, became known as Huntington's Chorea. The latter word, which comes from the Greek, and means dance, was used to describe the way in which the symptoms of the disease manifest themselves. In the less sensitive and un - enlightened days of the latter nineteenth century, that word was felt to be the most appropriate description of the symptoms of the illness. The ungainly movements of some who suffered from the disease, were equated with the movements of a wild sort of dance.

What thoughts does the word conjure up in the minds of most people? To most, the word "dancing" implies the opportunity to participate in an enjoyable experience. Graceful pirouettes of classical ballet; the high leaping athleticism of Cossack or Greek as they perform to the vibrance of a balalaika; the intricacies of a lively folk dance with its programmed sequence of steps, or the illusions of romance, as young or old take pleasure in the response of each other's bodies. In all types of dance, the dancers can take pleasure in the full control they exert over their own body movements. They can skip or leap; or turn and glide in response to the emotions of a musical rhythm.

The Forgotten Man of Lakeland

This sort of pleasurable experience gained through the co-ordination of mind and body could not be further away from the physical and mental conditions imposed on some people by Huntington's Disease, for there is nothing pleasurable in the manner in which this condition affects their lives. Huntington's Disease can ultimately cause a complete loss of control of body movements, as well as a deterioration in the functioning of a person's mind. Frighteningly so, it can also result in changed patterns of behaviour in a HD patient. The disease is still relatively uncommon, not only in Great Britain, but also throughout the world, even 120 years after George Huntington made his findings known.

HD is a genetic time bomb waiting to explode.

William Palmer must have inherited the disease from one of his parents. An explanation that is offered by some of his surviving, but distant relatives, is that in their opinion, their family's involvement with the disease, originated from a foreigner who was shipwrecked on the Cumbrian coast. As detailed medical records going back two or three hundred years are not available, and factual identification of the disease only goes as far back as 1872, descendants of William Palmer's immediate family accept that the origins of the disease in their family may even be put down to family folk lore.

In the nineteenth century, a person's life expectancy was considerably shorter than it is today, and some early members of William Palmer's family who may have carried the disease, possibly died from other causes before the symptoms of HD became apparent, or known. As his relatives discovered, HD is a disease which makes its effects felt throughout the whole family of a patient. Not only does it bring about a lack of physical control of limbs and bodily functions to the person affected, but it can also cause changes to, and a deterioration of, their mental and physical behaviour. This can create repercussions that can be felt throughout the family. Someone who has HD can become transformed into a completely different person.

Their behaviour can go through an extreme change; they become as unfamiliar to their families. One who was quiet and gentle can become aggressive, disruptive or even violent. When this occurs, some relatives find it extremely difficult to understand, and adapt to these changed circumstances. A husband is no longer the same person that a wife married, or a mother may become unrecognisable by her own children. The well loved personalities are no longer there; they are lost in the distorted mind or body of a stranger.

This is a brief background to the disease from which William Palmer died 82 years after the New York physician George Huntingdon first described it. It was established in 1983, by a research team headed by the Boston [USA] doctor, James Gusella, that it is chromosome 4 which carries the gene responsible for causing Huntington's Disease. This discovery was made almost thirty years after the death of William Palmer. In 1993, following ten years of intensive research, a remarkable breakthrough was made in finding and isolating the HD gene chromosome 4. Teams of scientists who were working simultaneously in America, Great Britain, Canada and Germany pooled their knowledge and expertise to arrive at this vital discovery. Yet although the location of the rogue gene has now been pin-pointed, there is still no known cure for the disease. But at least, isolating the gene is a major step forward. Diagnostic tests are now available that can establish whether a person is likely to be affected by the disease.

William Palmer may have been fortunate in that he lived through the greater part of his lifetime without being aware that he was carrying the disease. He was a physically fit, robust and active man who lived his life to the full, and there are no indications in his own writings or from family memories and information, that he suffered any chronic or major illness during his lifetime. The tragedy of William Palmer carrying the HD gene in his physical make up, was that, like many other carriers, he bequeathed the disease to his own children. It was the ultimate cause of death of his two daughters.

HD can be passed to each succeeding generation through the inherited genetic pattern from either parent. Each person inherits their own genetic blueprint from their parents, when the pattern of their physical and mental make up is set at the moment of conception. Their characteristics which include those of appearance, health and temperament are indelibly imprinted on the 46 chromosomes which are a fundamental part of our very existence. The chromosomes are arranged in pairs, one of each pair having come from each parent. In this way, half the chromosomes that a person inherits are taken from the father, while the other half is from the mother. It is therefore a one in two chance as to whether the HD affected chromosome is one of the 23 passed on by an affected parent.

Every chromosome contains a number of genes, each of which carries a substance which has a specific task to perform in the overall chemistry of a person's body. Just as the chromosomes carry genes essential to the well being of the integrated functioning of a body, they also can carry rogue genes. Therefore, if these are present, they may cause a breakdown in the complex system of physiology, which results in a malfunction of one, or more parts of the body system. The nature of the gene that causes HD is such that it is thought to cause changes in the operation of the body's mechanism by disturbing the chemical substances that affect the normal functioning of the brain.

Chromosomes are tiny and invisible to the naked eye, but they can be seen and studied with the aid of a powerful microscope. Over many years of study, scientists from different parts of the world who specialise in the complex study of genetics, have pooled their knowledge to evolve techniques whereby it is possible to determine the position and number of genes on a particular chromosome. Unlike some other diseases that are gender related, HD can be passed from any affected parent to any of their children. As HD is passed from one generation to another, there is the devastating certainty that if one parent carries the Huntington's gene in their own physical make up, there is a 50% chance that the disease will be passed on to their children.

It was William Palmer's son in law who first noticed there was something odd about his father in law's movements. William Palmer was normally a neat man, careful in his ways, and precise of manner. During the period of the second world war, Hamish Buchanan remained in Liverpool where he combined his daytime job as a chemist with the extra role of firefighter. He had left his family to live in the relatively safety of Kendal, to which he returned every fortnight. On one of these visits to the Palmer family home at Cliff Terrace. Hamish recalled that William Palmer's movements had become rather awkward. Although neither men were aware at that time of what was happening, this early sign fitted into the symptomatic pattern of the disease. One of the most common ways in which the disease becomes apparent is the onset of a slow, almost imperceptible deterioration of the physical and mental capabilities of a person. It creeps up on them, catching them unawares. It slows down a reaction here, or a response there. It happens in such a way that the changes in a person's well being or behaviour are hardly noticeable to either themselves or others. In many cases, the symptoms seem so slight and commonplace that there is a great danger of them being brushed aside, as merely being characteristic of the normal ageing process.

As one would expect of a rock climber and mountaineer, William Palmer was normally sure footed , but Hamish noticed that occasionally his father in law stumbled, and bumped into objects that were in his way. These little shows of clumsiness meant nothing at the time to Hamish, and they only assumed importance a number of years later when he saw similar symptoms manifested in the movements of his own wife Annie, William Palmer's eldest daughter.

Hamish became aware that his wife was awkwardly dragging one foot along the ground; she seemed unable to walk in a normal manner. For her, the lifting and stepping action of her right foot became impossible to achieve. The family sought medical advice, and eventually after many tests and consultations, it was diagnosed that Annie had Huntington's Chorea. By 1949, Annie Buchanan had become so severely disabled that she was unable to care for herself and needed all the help and support that her family could give. Eventually, even that was not enough, and she had to be transferred to a private nursing home which placed a great strain on the family's financial resources.

It was at about this same time, that it became obvious to more than Hamish that William Palmer was also affected by the disease, but it did not manifest with the same degree of rapidity and severity in the father, as it did in his daughter. While Annie needed full time supportive care, from her family and medical staff, William Palmer was still able to carry on with his work for almost another five years. During this period, his daughter Jean acted as his secretary and chauffeuse as he travelled about the country engaged in research work for his own books and revised versions of other guide books.

Jean was the Palmer's youngest daughter who gave up her own lifestyle to return to her parents home to help, and care for them. During the second world war, she had served in the ATS as a driver. She was proud of the fact that she was one of the team of instructors who taught Princess Elizabeth, now Queen Elizabeth II, the rudiments of driving and maintaining an ambulance. When Jean was demobbed at the end of the war, she went to live with her parents at their London home in Spencer Road, Clapham. For a time, she was able to resume her former career as a youth worker, where she ran a youth club in Paddington. Jean was able to successfully combine this with the job of being her father's secretary.

As first her mother became ill, and then HD began to seriously affect her father, Jean exchanged her role of chauffeuse for that of nurse, for she was the only member of the family capable of caring for both her parents. The time inevitably came when both William, and his wife Annie were unable to look after themselves in the extensive three storey house in which they lived. It was so large, that the top two floors were let off to other residents, while the ground floor was big enough for all the needs of the three members of the Palmer family.

Eventually Jean had to give up her job as a youth worker in order to take over the responsibility of caring for her parents on a full time basis. Her mother's latter years were plagued by blindness, and a chronic and severe sinus condition from which it was difficult for her to gain any relief. When Annie Palmer died in 1953, it proved a devastating loss to William for they had been devoted to each other during the long years of their married life.

After the death of his wife, William's own condition deteriorated rapidly, and Jean continued to care for her father at his London home for as long as she could. The strain began to tell, for the physical and emotional demands, imposed on her as the disease inevitably progressed, became too great. The situation worsened, the problems of patient care became too much for her to manage without help and Jean only gained some respite when William Palmer was admitted to hospital for a week or two. He was allowed home again as soon as Jean felt able to care for him. She found that nursing an HD patient is a demanding and exhausting task, but applied herself to the role with all that it involved. Information published by the Huntington Disease Association in their many leaflets that are readily available indicate some of the distressing aspects of the disease.

"Towards the end, things get much worse. They lose control of their facial muscles; they grimace; they can't swallow. Eating is difficult, and often the food has to be reduced to a pulp and fed through a tube so that they don't choke. They lose the power of speech, and are unable to talk, yet many people with the disease retain a clarity of mind, but are unable to communicate with other people."

"Most HD sufferers become incontinent, and are unable to control their bodily functions. Many of those who suffer from the disease have an insatiable appetite, and yet dramatically they lose weight. Saliva trickles from their mouths, as they lose control of facial muscles and they find it difficult to retain a normal sitting posture."

"Their uncontrolled body movements can become grotesque and frightening," explained a carer who regularly visited an HD patient in a Lancashire mental hospital. That particular patient was confined for the simple reason that forty years ago, there was no other place for her to receive care, and not a lot has changed in the provision of care for HD sufferers since then.

When a family is presented with such problems and difficulties, it can test the most caring, and loving of a patient's relatives. Weariness and the inevitability that whatever is done, will not bring about any improvement, coupled with a sense of guilt, can present real problems. Some relatives find they can not handle the situation through fear, and some have the awareness that even while they are caring for an HD parent, that disease may already be about to initiate their own slow, but inevitable journey towards death.

Ironically, while Jean was devoting herself to the care of her father, the beginnings of HD was already affecting her.

*William Palmer on
holiday in France
1949.*

*The debilitating
effects of
Huntington's
Disease become
apparent.*

Yet, in William Palmer's case, it could be argued that he was fortunate, if that word can ever be used in conjunction with such a devastating illness as Huntington's Disease. He had had the benefit of enjoying a full and active life for nearly all of his 77 years, and only succumbed to the debilitating effects of the disease during the last two years of his life.

The disease manifests itself in a number of different ways and usually takes a number of years to develop. While a very small percentage of known sufferers are under the age of 20, the majority of those affected show the symptoms much later in life. It is recognised that if an "at risk" person has not developed the symptoms by the age of 60, then there is a very strong possibility that the gene has not, in fact, been inherited after all. Although it has been established that the onset of the disease can appear at any age, the older an "at risk" person becomes, without the symptoms appearing, the less chance there is of succumbing to the disease. As the years progress and no alarming signs appear, there is a strong possibility that a person "at risk" may have escaped the disease, if not the years of strain and tension wondering when the symptoms would appear.

William Palmer escaped those ten to fifteen years of illness that affects most HD sufferers, where the degree of increasing severity varies with each individual. In William Palmer's case, the effects brought about by the disease were obviously not so severe in the years leading up to his death, as those experienced by others, for they certainly did not prevent him from carrying on with his work. His grand - daughter recalled that her grandfather was able to work almost up to the end of his life.

There are periods when the disease appears to go into remission, but this is an illusion only. Symptoms return and manifest themselves in different degrees of severity, as the disease passes through various stages of advancement Inevitably, those who suffer from HD have their ups and downs physically, emotionally and mentally. Some days, hopes may be raised as patients appear to show improvement, while at other times they seem to be dramatically worse. When the good days occur, it is merely an illusion of well being for family and friends, for unlike a prisoner fastened in a cell, the HD patient is unable to obtain remission for good behaviour. Locked within a deteriorating frame, the spirit seeks to escape the body's own imprisoning cells. These periods of temporary stability are double edged, for while the relative calm of a seemingly improved situation may be regarded as a breathing space for both patient and carer, there is the inevitable realisation that the next stage will be one more closer to the end.

A photograph of William Palmer taken while on holiday in France in 1949, shows him to be casual and relaxed, while another photograph taken at some later, but of an unknown date, shows the strain and tension of the illness etched in every line of his face.

Periods of home and hospital care for William Palmer followed a steady pattern, through the months of 1954, and after a further period of home care late in the year, William Palmer was once again admitted to hospital just before Christmas. Jean was advised by the medical staff and her own doctor to use that opportunity to have a holiday herself, as the condition of her own health was beginning to cause concern. In the same way, as HD sufferers show different levels of the disease, so there are carers who have a higher tolerance level to face up to the strains imposed on them. Some can continue to care for a loved one in their own homes, far beyond the expectations of what one would normally expect to be possible. Others appreciate the expertise, comfort and relief that professional care can offer.

Jean decided to travel north, to spend the Christmas period with her sister's family who were then living in Heysham, in Lancashire. The journey in the dark, dismal days of winter was difficult and depressing. With a rail system that was still recovering from the effects of the second world war, delays and diversions were commonplace. It was as a result of such difficulties that Jean's train was sent to Carlisle instead of Lancaster, and she eventually arrived at her sister's home several hours later than anticipated; tired and exhausted.

Jean's planned holiday failed to materialise, for shortly after her arrival, a phone call from Paddington Hospital informed her that William Palmer's condition had deteriorated rapidly. There was nothing to do but return south as soon as possible. Jean made the weary return journey to London by the next available train. She was accompanied by her niece, William Palmer's only grand daughter.

It was Christmas Eve when they reached London, and Christine Buchanan recalled the final moments of her grandfather's life. The end came on the 26th December after William Palmer had spent the last four months in and out of hospital.

"At the end of his life his speech became very blurred, but as far as I can remember he could make himself understood, at least to Jean. I'm afraid I found him rather frightening at that time. He could still walk at the end of his life but only with help. In hospital, he developed pneumonia however and died - very peacefully; Jean and I were beside him."

His granddaughter suggested that the ultimate cause of his death was a combination of three factors, old age, Huntington's disease, and a broken heart caused by the death of his wife to whom he had been happily married for over fifty years.

William Palmer was buried, in a grave that he shared with his wife, among the city smoke of London. It lies hundreds of miles away from his native fells, that shelter the freshness of the valleys, lakes and sparkling streams of an area that he loved so much.

There is no known epitaph, or monument to remind present generations of what William Palmer did, or what he was. Nowhere is there any tangible acknowledgement of the success that he showed it was possible to achieve, for one who came from humble beginnings and background. Only the rows of old time bindings on library shelves bear testimony to the pleasure he gave to others, through his writing.

Over many years his memory, like his bones, has laid undisturbed.

"William Palmer, the Forgotten Man of Lakeland."

Borrowdale birches.

Appendix A - Camping Equipment

This is the suggested list of camping equipment that William Palmer recommended for two persons, with the appropriate weights given in pounds and ounces.

		lbs	ozs
1.	Tent: Itisa	1	13
2.	Flysheet	1	00
3.	Ground sheet	1	11½
4.	Ground blanket		11
5.	Sectional poles	1	12
6	Skewer tent pegs		1¼
7	Tent peg Bag for knife fork and spoon		2
8	Pole discs		2
9	Pole pocket sewn to rucksack		¼
10	Bergen type rucksack	1	2
11	Primus stove half full	1	10
12	Meta tablets		5
13	Frying pan		2½
14	Plate		½
15	2 pans		8
16	2 pots for jam and butter		½
17	waterproof tea bag		¼
18	waterproof bags for cocoa, sugar and salt		½
19	bread bag and tea cloth		1
20	tea infuser, muslin		¼
21	waterproof bag for oatmeal		½

To these were added the following items that made up his personal equipment.

		lbs	ozs
1	Enamelled cup		3
2	Pot handle		1
3	Mirror		1
4	Towel		6
5	Toothbrush and paste		1½
6	Knife,fork & two spoons		3
7	Candle		1
8	Toilet & shaving soap		1
9	Shaving brush		½
10	Night attire		9
11	Extra pair of socks		3
12	Housewife repair kit, needles etc.		3½
13	Swimming costume		2
14	First aid kit		4

Appendix B

This is William Palmer's cycle route for "Girdling the Lake District", about which he wrote in Wanderings in Lakeland.

As it is a circular route, the start may be made at any point on the "Girdle." When William Palmer planned the route, it was by starting from his home town of Kendal. It seems logical therefore, to use that as the starting point for this description.

Road conditions have altered in the sixty years since William Palmer planned this route. It seems unlikely that he would ever have contemplated the volume of traffic that now uses many of his recommended roads. Tourist attractions have also been developed to meet the needs of the millions that visit the Lake District throughout the year. Many of these were not in existence when he described the route.

While I have attempted to keep to William Palmer's original route, optional alternatives are suggested in *italics* and places to visit are suggested in **[bold brackets]**.

Although William Palmer suggested this as a route for cyclists, most of it would be easily and equally possible in a car, and would form the basis for a week's touring holiday of the Lake District.

Day 1

Leave Kendal for the Ferry that crosses the south end of Windermere. Then travel by Sawrey, **[opportunity to visit Hilltop, home of Beatrix Potter]**, before proceeding via Hawkeshead **[Grammar School attended by Wordsworth]** for Coniston. **[Ruskin Museum, Brantwood and Steam Yacht Gondola]**

A choice of route can be taken at Coniston. The strenuous option is to walk and push the cycle over the Walna Scar track **[opportunity to see old mines and slate quarries]** which leads to Seathwaite in the Duddon Valley. The main route follows the road through Torver to Broughton in Furness where accommodation is available.

Day 2

Leave Broughton in Furness for Duddon Bridge from where a steep road heads north east towards Ulpha. The rider has a choice of routes for the left fork offers easier riding that leads to Boot in Eskdale. **[Dalegarth Falls]** while the right fork takes the rider into the heart of mountain scenery. The next village on this route is Seathwaite **[chapel where the Wonderful Walker was vicar]**. The road follows the Duddon **[look out for the beautiful Birks Bridge]** before reaching Cockley Beck **[site of an ancient** monument, Thingmount]. Turn west over Hardknott Pass, [**Hardknott Roman Fort**] walking and pushing where necessary before descending into Eskdale.

There are interesting grave stones in [**St. Catherines churchyard**], and the [**old mill**] at Boot is worth a visit. Take time to visit Dalegarth Station to see the [**Ratty Railway**]. *Leave Eskdale by riding west to spend a night in the small seaside port of Ravenglass* [**Roman bath house & nearby Muncaster Castle**] .

Camping and accommodation is available

Day 3

Transport for walkers and cyclists is available from Ravenglass station to the [**Sellafield Visitors centre**]. *For riders with time to spare, this is an interesting option.* [**Muncaster Castle**] *is also nearby. The castle and gardens, and the fascinating Owl centre make it well worth a visit. Either of these options is likely to take a full day out of the programme, but may make a welcome break from days in the saddle.*

Leave Ravenglass for Holmrook turning east towards Irton . Follow the road to Santon Bridge, where a fell road leads to Nether Wasdale. Take the road that parallels Wastwater almost all the way to Wasdale Head. [**The inn is that used by the old time climbers, tiny church, magnificent mountain scenery**]. The return must be made along the same road as far as the junction with the road to Gosforth. Turn right for some up and down riding before Gosforth is reached. [**Viking cross in churchyard**]. Join the main A595 road as far as Calderbridge, turn right by the church to take the fell road. [**Calder Abbey**]. The road is steep in parts, be prepared to push, and beware of Sellafield's speedy commuters. Follow the road down to Ennerdale Bridge, for accommodation and camping.

Day 4.

Leave Ennerdale Bridge by the initially steep road to Lamplugh. Pass the church and almost immediately turn right to follow the fellside road to Fangs Brow. A steep descent brings the rider alongside Loweswater before crossing the river Cocker at the foot of Scale Hill. *A bridle way through the woods is a short cut option that brings the rider out on the valley road part way alongside Crummock.* Non - woodland riders will need to follow the road towards Lorton and take the turning on the right for Buttermere. The cyclist pedals along the eastern shore of Crummock for the hamlet of Buttermere.[interesting little church with memorial to Alfred Wainwright.] Refreshments, accommodation and camping all available. *Route options, turn north east and follow the road to Keswick via Newlands pass. Steep climbs involve pushing and walking on both sides of the pass.*

The main route carries on into the heart of the Buttermere valley to Honister Pass [old **slate quarries**]. Pushing and walking for all but the hardest rider, with very steep descents into Seatoller in Borrowdale [**Visitor Centre**]. Relatively easy ride alongside Derwentwater to reach Keswick with plenty of accommodation and facilities for camping.

Day 5

William Palmer's route from Keswick [**museums of pencils, cars, musical stones**] would take the rider along the now busy A66 past Threlkeld, and Blencathra to the former site of sheep pens at Troutbeck where the route turns right for Ullswater.

This can be avoided by following a bye road out of Keswick for the [**Castlerigg Stone Circle**]. *Continue along the road that passes the circle towards St. Johns in the Vale. A mini cross road just past Wanthwaite Bridge will give access to an old coach road. It's rough surface offers a quieter alternative to the main road, but rejoins William Palmer's route at Dockray.* From there, the road descends to Ullswater [**Aira Force**], and the northern side of the lake can be followed to Pooley Bridge.

William Palmer's recommended route is to visit Haweswater by going through the village of Bampton, before making the return to Kendal by way of Shap Fell. Shap rises to over 1000 feet, but the surface of the road is good, and the A6 is now a relatively quiet road since the opening of the M6.

Before the rider reaches Kendal, a right turn indicated to Burneside will take one through William Palmer's home territory. Turn right in Burneside for Bowston to see [**the cottage where William Palmer was born**]. *The return journey through Burneside will complete the journey to Kendal.*

If the rider has time available, William Palmer's route can be adjusted, to enable riders to visit extra lakes by the taking the following options.

After visiting Haweswater, instead of heading south on the A6, as William Palmer suggested, the rider should return to Pooley Bridge where there is plenty of accommodation and camping facilities.

Day 6

Leave Pooley Bridge to follow the northern shore of the lake to the end of the lake at Glenridding. Steep work ahead in the shape of Kirkstone Pass with Brotherswater within easy range. Walking and pushing is needed by most riders to reach the top of the pass with its welcoming inn. A short way over the summit, the road forks to the right for the steep descent to the waiting fleshpots of Ambleside.

Turn right on the A595 for Rydal **[Rydal Mount]** *and Grasmere* **[Dove Cottage].** *There is plenty of accommodation, and the possibility of an energetic option of a night time ride. Leaving Grasmere by the main road over Dunmail Raise, Thirlmere, formerly known as Leathes Water is circled by a road, and provides an additional lake to be ticked off by a fit rider before returning to base at Grasmere.*

Day 7

Leave Grasmere either by retracing yesterdays route to Rydal, and then turn right over Pelter Bridge to follow the river Rothay as far as Rothay bridge before resuming the main road at Waterhead. The road alongside Windermere is followed **[Brockhole Visitor's centre]**, *before turning right for Bowness.* **[The World of Beatrix Potter]** *The return to Kendal can be made by the fell road to Crook. Turn left for Staveley and follow the road to Kendal by way of William Palmer's home territory of Bowstone and Burneside for the return to Kendal.*

*Girdling the Lakes -
a route for hard riders.*

Appendix C - Sequence of events in the life of William Palmer

1877	8th July, born at 4 Bowston Terrace, Burneside.
1881	Census records included both parents and their five children.
1886	First visit to Grasmere Sports
1890	Left school
1894	Long walk of 85 miles in 24 1/2 hours aged 17
1895	Attempt on 4 peaks record with Messrs Dawson and Poole. Then aged 18
1901	30th Sept married Annie Ion
1902	Lake Country Rambles first book published -Chatto & Windus
1903	Lakeland Dells and Fells - Chatto & Windus
1905	English Lakes A & C Black, illust. WH Cooper
1905	Recorded in Bulmer as living at 5 Lake Road Terrace, Kendal occupation given as a compositor
1905	Exploratory work for 1st ascent NW route Pillar, with Fred Botterill
1905	Daughter Annie born
1906/15	Lived at 4 Beechwood, Kendal
1906	Joined Fell and Rock as an original member.
1908	English Lakes [reprinted]
1910/18	Editor of Fell and Rock Journal
1912	Daughter Jean born
1912	Records in editorial notes of Journal, delay in publication due to great difficulties with eyesight, and personal problems
1913	English Lakes [reprinted twice]
1913	Odd Corners of English Lakeland - Skeffington
1914	Odd Yarns of the English Lakes -Skeffington
1914/18	Rejected for military service because of eyesight problems. Became recruiting officer for Kendal

1916   Suffered a breakdown while writing editorial notes for Journal. Completed by his wife.

1917/8 Joint editorship of Journal with wife.

1918   English Lakes [reprinted]

1918   Became a life member of the Fell and Rock Climbing Club

1918/33 Recorded as living at 69 Salisbury Road, Liverpool

1921   Elected a member of the Wayfarers Club, Liverpool.

1925   English Lakes New Edition

1926   Things Seen at the English Lakes - Blacks

1926   His daughter, Miss A Palmer is elected as a Life member of F & R CC

1929   English Lakes New edition [Reprinted]

1930   English Lakes [Kitbag] - George Harrap & Son

1930   10th January became FRGS

1932   Wales Its History and Romance- Kitbag; Harrap

1934/39 Lived at High Wray, Ambleside

1934   The Complete Hill Walker - Pitman

1934   Tramping in Lakeland - Country Life

1934   Tramping in Derbyshire - Country Life

1936   English Lakes New Edition [reprint]

1937   Odd Corners of the Yorkshire Dales [1st Ed] - Skeffington

1937   Odd Corners in North Wales [1st Ed]-Skeffington

1937   More Odd Corners of English Lakeland [1st Ed] Skeffington

1938   Verge of Lakeland - Robert Hale

1938   Odd Corners of Derbyshire [1st Ed] - Skeffington

1939/45 7 Cliff Terrace Kendal

1939   Elected member of the British Ornithologists Union

1939   Penguin Guide to the Lake District

1939   Verge of Scotland - Robert Hale

1941   Verge of Western Lakeland - Robert Hale

1943   December. Elected a Fellow of the Society of Antiquaries of Scotland

1944	Verge of Wales - Robert Hale
1944	The River Mersey - Robert Hale
1944	Odd Corners in North Wales [2nd Ed]
1944	More Odd Corners in English Lakeland [2nd Ed]
1944	More Odd Corners in North Wales
1944	Odd Corners in the Yorkshire Dales - Skeffington
1944	[Feb] - Odd Corners in English Lakeland [2nd Ed]
1944	[Feb] - Odd Corners of Derbyshire [2nd ed]
1946/52	5 Earlsfield Road, Wandsworth Common,London. There is a reliable suggestion that William Palmer was revising Ward Lock guide books during these years
1946	Wanderings in Lakeland -Skeffington war economy
1946	Wanderings in Scotland - Skeffington
1946	[June] - Odd Corners in North Wales [3rd Ed]
1946	[June] - Odd Corners in English Lakeland [3rd Ed]
1946	[July] - More odd corners in English Lakeland [3rd Ed]
1946	[July] - Odd Corners of Derbyshire [3rd Ed]
1946	[Dec] - Odd Corners in North Wales [4th Ed]
1946	[Dec] - Odd Corners in English Lakeland [4th Ed]
1946	[Dec] - More Odd Corners in English Lakeland [4th Ed]
1946	[Dec] - Odd Corners of Derbyshire [4th Ed]
1947	Attended Grasmere Sports to see his great nephew win youth's guide race.
1947	Verge of the Scottish Highlands - Robert Hale
1948	[Feb] - Odd Corners in North Wales [5th Ed]
1948	Odd Corners in Derbyshire [5th Ed]
1948	[Feb] - Odd Corners in English Lakeland [5th Ed]
1948	[Feb] - More Odd Corners in English Lakeland [5th Ed]
1948	[Feb] - Odd Corners of Derbyshire [5th ed]
1949	Natures Calendar - Skeffington
1949	Penguin Guide to North Wales

1950[?] Wanderings in Surrey - Skeffington

1951   Wandering in the Pennines - Skeffington

1951   Wanderings in Ribblesdale - Skeffington

1951   Odd Corners in Surrey - Skeffington

1952/4   Living at 2 Spencer Road London SW18

1952   2nd edition Penguin Guide to North Wales

1952   Byeways in Lakeland - Robert Hale

1953   His wife, Annie Palmer died

1954   William Palmer died, 26th December.

The following book has no dated publication;

The Art of Camping - published by Witherby

Family camping holiday 1951.

Appendix D - Huntington's Disease

The HD disease is to be found among people in many different parts of the world; it is not confined to those of a particular race or culture. It is currently estimated that 6,000 people in the United Kingdom have Huntington's Disease, with a further 50,000 at high risk. As the symptoms of HD are so similar to other diseases that attack the body's central nervous system, this number could be an under estimate of the true scale of the problem.

The birth of children to HD carriers is obviously the way of increasing the number of people affected, and this poses its own difficult decision making. Many couples, who know there is a high risk of passing on the disease to any children they may have, are faced with deciding whether they have a moral right to impose the possibility of their burden on an unborn child. Others, conscious of how the disease has affected their own lives, opt for a childless relationship.

The current known figure of 6,000 people that are actually affected by HD may seem a small number when set against the millions that inhabit these islands, especially when compared to the ratio of other terminal diseases. But for a person who has the misfortune to be one of that 6,000, the problems they and their families have to face are out of all proportion to the actual numbers affected.

As greater advances in knowledge and awareness of the disease have become available, a test has been developed that has a fair degree of accuracy, which can diagnose the presence of the gene in "high risk" people. The nature of the disease has been termed a form of genetic Russian roulette, for even though a person may be aware that there is a 1 in 2 chance of being affected by the disease, conversely there is a 1 in 2 chance that they will escape, yet they go through their lives with the uncertainty of HD hanging over them.

The subsequent results of the tests that are now available can promote different reactions in the people concerned. It was only natural that one young woman was elated by the results of her test which gave her the "All Clear." When the suggestion was made to her that this would come as a great relief from the tension and worry of waiting for the worst to happen, her reply was philosophical and sobering. "When you've grown up with the knowledge that this disease is in the family, and attended the funerals of ten relatives that have died as a result of HD , you come to terms with it and just accept what comes."

Another person was almost reassured to be told that her staggered way of walking was due to the illness of HD , even though she knew it was one for which there was, as yet, no known cure. The diagnosis put an end to the taunting that her children were receiving at school, through being told that their mother was always drunk.

Lack of awareness, and knowledge by the public about the disease, can lead to these misconceptions. There can be a build up of frustration on the part of carer or patient, caused through the increasing difficulty of making ones feelings and wants known. The basic systems of communication with others breaks down; isolation becomes more of a reality. Often this leads to temper tantrums, which can result in violent and irrational behaviour. The tragedy is that inside the uncontrollable shell of a HD body, is a person, desperate to be understood.

Help in understanding the needs of HD patients and their families, and in coming to terms with the disease can be obtained through the services provided by the Huntington's Disease Association. Self help groups have also developed in many different parts of the country and provide a network of support through information, expertise, counselling and encouragement to HD patients and their families.

The HDA publishes a number of leaflets and information documents on the disease. These are available from The Huntington's Disease Association, 108 Battersea High Street, London SW11 3HP.